SpringerBriefs in Computer Science

More information about this series at http://www.springer.com/series/10028

Denise Demirel • Lucas Schabhüser
Johannes Buchmann

Privately and Publicly Verifiable Computing Techniques

A Survey

Springer

Denise Demirel
Theoretische Informatik
Technische Universität Darmstadt
Darmstadt, Hessen, Germany

Lucas Schabhüser
Theoretische Informatik
Technische Universität Darmstadt
Darmstadt, Hessen, Germany

Johannes Buchmann
Theoretische Informatik
Technische Universität Darmstadt
Darmstadt, Hessen, Germany

ISSN 2191-5768   ISSN 2191-5776   (electronic)
SpringerBriefs in Computer Science
ISBN 978-3-319-53797-9   ISBN 978-3-319-53798-6   (eBook)
DOI 10.1007/978-3-319-53798-6

Library of Congress Control Number: 2017931973

Printed on acid-free paper

This Springer imprint is published by Springer Nature
The registered company is Springer International Publishing AG
The registered company address is: Gewerbestrasse 11, 6330 Cham, Switzerland

Preface

Verifiable computing refers to methods that allow delegating the computation of a function on outsourced data to a server, such that the data owner and/or third parties can verify that the result has been computed correctly. Those approaches are even more useful when they provide a verification process that is more efficient than performing the computation locally. To address this challenge, many techniques have been proposed. In this work, the first comprehensive survey of all existing constructions is provided. In doing so, we are concerned with a setting where three parties are involved: A *client* who provides some input data, a *server* who evaluates a function on the input data, and a *verifier* who verifies the correctness of the result. Schemes dealing with a more complicated setting of multiple clients, verifiers, or servers are beyond the scope of this work. Furthermore, we do not consider approaches that rely on replication, trusted hardware, remote attestation, or spot checking.

For all approaches that match our setting and allow for a sufficiently efficient verification process, we provide a brief description of the approach and highlight the properties the solution achieve. More precisely, we analyze which level of security it provides, how efficient the verification process is, whether anyone or only the client can check the correctness of the result, which function class the verifiable computing scheme supports, and whether privacy with respect to the input and/or output data is given. Based on this analysis we compare the different approaches and outline possible directions for future work.

Darmstadt, Germany
June 2016

Denise Demirel
Lucas Schabhüser
Johannes Buchmann

Acknowledgments

This work has been co-funded by the European Union's Horizon 2020 research and innovation program under Grant Agreement No 644962. In addition, it has received funding from the DFG as part of project "Long-Term Secure Archiving" within the CRC 1119 CROSSING. The authors wish to thank Daniel Slamanig and David Derler for their valuable input and helpful discussion during the writing of this work.

Contents

Chapter 1
Introduction

Abstract In this chapter we provide an introduction to our survey. First, we motivate the research field of verifiable computing. Afterwards, we present a roadmap to this book explaining the setting we a concerned with followed by the relevant properties, i.e. the level of security, privacy, and efficiency, the type of verifiability, and the function class provided. Finally, we show the organization of our work.

1.1 Motivation

Due to the increasing popularity and relevance of cloud computing there is an increasing market for solutions that allow to outsource data and computations to the cloud. However, since the servers performing these computations might be malicious or infected (and can thus not be fully trusted), it is a desirable feature that clients or third parties can verify the correctness of any outsourced computation. A naïve solution to this problem is to reobtain the outsourced data from the cloud, check its integrity, and reexecute the computation locally. However, in this case the verifier, i.e. the client or a third party, needs to have enough space to store the data and enough power to perform the computation. This is not a viable solution especially for weak devices, such as smartphones, huge amounts of data, and time-consuming computations. Moreover, it negates many of the benefits of using the cloud in the first place. Thus, a vital research question is how the verification of the correctness of a computation can be performed while *requiring less computational work* than a local computation and optimally without needing access to the input data. The field of *verifiable computing* aims at providing solutions to this problem.

In 2009, Genarro et al. [3] provided the first definition of a non-interactive *verifiable computing* scheme and since then many solutions for different types of computations using vastly different approaches have been presented. As this field has significantly grown in the last few years and is still growing quite fast, this work surveys the current state of the art for different types of verifiable computing schemes. The existing constructions are described and their properties, e.g. level of security, efficiency, and privacy, are outlined and compared. To conclude, based on this comparison possible directions for future work are outlined.

© The Author(s) 2017

D. Demirel et al., *Privately and Publicly Verifiable Computing Techniques*,
SpringerBriefs in Computer Science, DOI 10.1007/978-3-319-53798-6_1

1.2   Roadmap

In this work we will be concerned with a setting where three parties are involved: a *client* who provides some input data x, a *server* who evaluates a function f on input data x, and a *verifier* who verifies the correctness of result y, i.e. whether $y = f(x)$ holds. If the server indeed computed y correctly, then the verifier should accept the result and otherwise reject it with high probability. In order to establish this correctness guarantee, either the verifier interactively asks the server questions about the computation performed by the server or the server returns audit data to the verifier which allows it to locally check the correctness of the computation. This work considers the predominant model containing a single client, a single verifier, and a single server. Works dealing with a more complicated setting of multiple clients, verifiers, or servers, that have for instance been proposed in [2], are beyond the scope of this work. This survey aims at providing an overview of state of the art verifiable computing. There exists substantial prior work in this field (see for example [1]) which is not listed in this survey.

Now, let us briefly informally introduce the most important properties of verifiable computing schemes, i.e. the level of security, privacy, and efficiency, the type of verifiability, and the function class provided.

Security.   A verifiable computing scheme is secure, if a malicious server will not be able to convince a verifier of the correctness of a computation although the result is not correct. The security level provided describes the capability of an adversary when attacking the scheme and will be formally defined in Sect. 2.1.

Privacy.   Privacy comes in different flavours. It can be related to the servers' or the verifiers' point of view. In both cases privacy means that the parties do not learn the input data, the result, or even the input *and* result of a computation. Such a feature is desirable when dealing with sensitive data, e.g. medical data, government data, business data, and military data. Think, for instance, of a cloud storing medical records and performing statistics on them. In this scenario the cloud and the verifiers should not learn the medical condition of individual patients while processing and verifying the data.

Efficiency.   Efficiency considers the work required by the client and the verifier.[1] Usually the client has to perform some preprocessing which is an additional input to the computations performed by the server. If both the preprocessing and the verification are more efficient than performing the computation locally a verifiable computing scheme is called to be efficient. However, efficiency is often considered in an amortized sense, i.e. the client has some setup costs which are performed once and then multiple operations can be performed and verified. Thus, schemes where at least the verification process is more efficient than performing the computations are said to provide amortized efficiency.

---

[1]To analyse and improve verifiable computing schemes with respect to the effort spent by the server is also an interesting research direction. However, so far this aspect has not been rigorously covered in literature which is why we also do not address it here.

Public/private verifiability. There are two types of verifiability that are distinguished. The verification can either be performed by any third party (public verifiability) or requires some secret information from the client (private verifiability).

Function class. The function class provided determines the expressiveness of the computations that can be handled by the verifiable computing scheme. While some schemes cover only arithmetic circuits of fixed degree others can handle arbitrary arithmetic circuits or even arbitrary C code.

In this work we do not consider approaches that rely on replication, i.e. that outsource the same computation to n independent servers and use majority voting on the results to determine the correctness. Since this assumes uncorrelated failures, this assumption seems too strong. Also, we do not consider the use of trusted hardware on the server side, remote attestation, or spot-checking. The former two approaches require trusted hardware assumptions and additionally trusted hardware is usually strongly limited in scalability. The latter approach relies on the assumption that failures, if they occur, are very frequent, which also seems to be a quite strong assumption.

1.3 Organisation

We first provide some general definitions in Chap. 2. Afterwards, in Chap. 3 we present the *proof* and *argument* based systems that have been developed in recent years and for which a variety of tools are already available. Chapter 4 covers schemes based on *fully homomorphic encryption* and in Chap. 5 we survey schemes based on *homomorphic authenticators* (message authentication codes and signatures). Chapter 6 covers solutions from functional cryptography, while in Chap. 7 individual schemes that allow for the verification of specific computations are presented. Finally, Chap. 8 provides a summary and comprehensive analysis of the different types of verifiable computing schemes followed by a conclusion and possible future work in Chap. 9.

References

1. L. Babai, Trading group theory for randomness, in *Proceedings of the 17th Annual ACM Symposium on Theory of Computing*, Providence, RI, 6–8 May 1985, pp. 421–429
2. D. Fiore, A. Mitrokotsa, L. Nizzardo, E. Pagnin, Multi-key homomorphic authenticators, in *Advances in Cryptology - ASIACRYPT 2016 - 22nd International Conference on the Theory and Application of Cryptology and Information Security, Proceedings, Part II*, Hanoi, 4–8 December 2016, pp. 499–530
3. R. Gennaro, C. Gentry, B. Parno, Non-interactive verifiable computing: outsourcing computation to untrusted workers, in *Advances in Cryptology - CRYPTO 2010, 30th Annual Cryptology Conference, Proceedings*, Santa Barbara, CA, 15–19 August 2010, pp. 465–482

Chapter 2
Preliminaries

Abstract In this chapter we provide formal definitions for verifiable computing schemes and their relevant properties. More precisely, first, we define *verifiable computing schemes* in general and *privately verifiable computing schemes* and *publicly verifiable computing schemes* in particular. Then, we provide a definition for *weak* and *adaptive security*. Following, we discuss the different types of privacy protection, i.e. *input privacy w.r.t. the server*, *input privacy w.r.t. the verifier*, *output privacy w.r.t. the server*, *output privacy w.r.t. the verifier* and give a definition for each property. Finally, we define efficiency distinguishing between *efficiency* and *amortized efficiency*. Many verifiable computing schemes presented in the subsequent chapters are constructed with the help of cryptographic primitives that come with additional definitions for the underlying hardness assumptions. However, since these are very specific to the individual solutions they are presented in Appendix A.

2.1 Verifiable Computation

In this work we will always consider the following scenario. A client C provides some input x to a server S. Then S is asked to evaluate a (where appropriate encoded) function f on input x. S will do the computation and then send the result y to a verifier V. To prove the correctness of the result to V, i.e. to prove that y is indeed equal to $f(x)$, a verifiable computing scheme can be used. In the following we use the definition of a non-interactive verifiable computing scheme introduced by Gennaro et al. [2].

Definition 2.1 (Verifiable Computing Scheme) A *Verifiable Computing Scheme* VC is a tuple of the following probabilistic polynomial-time (PPT) algorithms:

KeyGen($1^\lambda, f$) : The probabilistic key generation algorithm takes a security parameter λ and the description of a function f. It generates a secret key sk, a corresponding verification key vk, and a public evaluation key ek (that encodes the target function f) and returns all these keys.

ProbGen(sk, x) : The problem generation algorithm takes a secret key sk and data x. It outputs a public value σ_x which encodes the data x and a corresponding decoding value ρ_x.

D. Demirel et al., *Privately and Publicly Verifiable Computing Techniques*,
SpringerBriefs in Computer Science, DOI 10.1007/978-3-319-53798-6_2

Compute(ek, σ_x) :   The computation algorithm takes the evaluation key ek and
the encoded input σ_x. It outputs an encoded version σ_y of the function's output
$y = f(x)$.

Verify(vk, ρ_x, σ_y) :   The verification algorithm obtains a verification key vk and
the decoding value ρ_x. It converts the encoded output σ_y into the output of the
function y. If $y = f(x)$ holds, it returns y or outputs $\perp$ indicating that σ_y does not
represent a valid output of f on x.

Definition 2.2 (Correctness) A verifiable computing scheme VC is correct if for
any choice of f and output (sk, vk, ek) $\leftarrow$ KeyGen($1^\lambda, f$) of the key generation
algorithm it holds that $\forall\, x \in$ Domain(f), if (σ_x, ρ_x) $\leftarrow$ ProbGen(sk, x) and $y \leftarrow$
Compute(ek, σ_x), then $y = f(x) \leftarrow$ Verify(vk, ρ_x, σ_y).

In the original work on non-interactive verifiable computing Gennaro et al. [2]
only considered privately verifiable computing schemes as defined below.

Definition 2.3 (Privately Verifiable Computing Scheme) If sk = vk and C needs
to keep ρ_x private, VC is called a *privately verifiable computing scheme*.

Clearly, such a scheme requires the client to run the verification algorithm. Later in
[3], Parno et al. introduced the notion of publicly verifiable computing schemes.

Definition 2.4 (Publicly Verifiable Computing Scheme) If sk $\neq$ vk, VC is called
a *publicly verifiable computing scheme*.

It allows to hand out vk to third parties without revealing sk. Therefore, everyone
with knowledge of vk and ρ_x can verify the correctness of the server's computation.

Intuitively the difference between the two notions is that in privately verifiable
computing the client keeps its verification key secret. It follows that only the client
can act as verifier. Note that in privately verifiable computing schemes revealing the
verification key often leads to a loss of security. More precisely, knowledge of the
verification key allows the server to compute a wrong result leading to a correct
verification proof. In publicly verifiable computing, on the other hand, knowledge
of the verification key does not help a malicious server to forge an incorrect result.
Thus, it can be published allowing not only the client but anyone to act as verifier
and to check the correctness of a performed computation.

2.2 Properties of Verifiable Computing Schemes

In this section a definition for security, privacy, and efficiency is given. We will
mainly follow the approach of Gennaro et al. [2], who were the first to define
verifiable computing schemes. In addition, we also integrate some later proposals
to obtain stronger security definitions, e.g. adaptive security presented in [1].

2.2.1 Security

Intuitively a verifiable computing scheme *VC* is secure, if a malicious server cannot persuade the verification algorithm to output $y^* \neq f(x)$ except with negligible probability. Formally, we define the following two experiments. We distinguish between two types of adversaries, a weak adversary and an adaptive adversary. The weak adversary [2] only has oracle access to ProbGen but is not allowed to call Verify in the privately verifiable computing setting. It can only try once to have an incorrect result verified as correct but must never learn the client's acceptance bit, since this information might be used to produce subsequent forgeries. An adaptive adversary [1] can run $\mathbf{EXP}_A^{\text{Verify}}$ multiple times, by calling $\mathbf{EXPadapt}_A^{\text{Verify}}$, and learn about the client's acceptance bit and adapt its forgeries accordingly.

Experiment $\mathbf{EXP}_A^{\text{Verify}}[VC, f, \lambda]$:
  $(\text{sk}, \text{vk}, \text{ek}) \leftarrow \text{KeyGen}(1^\lambda, f)$
  for $i = 1, \ldots, \ell = \text{poly}(\lambda)$ **do**
    $x_i \leftarrow A(\text{ek}, x_1, \ldots x_{i-1}, \sigma_1, \ldots, \sigma_{i-1})$
    $(\sigma_i, \rho_i) \leftarrow \text{ProbGen}(\text{sk}, x_i)$
  end for
  $(i, \sigma_y^*) \leftarrow A(\text{ek}, x_1, \ldots, x_\ell, \sigma_1, \ldots, \sigma_\ell)$
  $y^* \leftarrow \text{Verify}(\text{vk}, \rho_i, \sigma_y^*)$
  if $y^* \neq \bot \wedge y^* \neq f(x)$ **then**
    return 1
  else
    return 0
  end if

Experiment $\mathbf{EXP}_A^{\text{AdaptVerify}}[VC, f, \lambda]$:
  $(\text{sk}, \text{vk}, \text{ek}) \leftarrow \text{KeyGen}(f, 1^\lambda)$
  for $j = 1, \ldots, m = \text{poly}(\lambda)$ **do**
    for $i = 1, \ldots, \ell = \text{poly}(\lambda)$ **do**
      $x_i \leftarrow A(\text{ek}, x_1, \ldots x_{i-1}, \sigma_1, \ldots, \sigma_{i-1}, \delta_1, \ldots, \delta_{j-1})$
      $(\sigma_i, \rho_i) \leftarrow \text{ProbGen}(\text{sk}, x_i)$
    end for
    $(i, \sigma_y^*) \leftarrow A(\text{ek}, x_1, \ldots, x_\ell, \sigma_1, \ldots, \sigma_\ell, \delta_1, \ldots, \delta_{j-1})$
    $y^* \leftarrow \text{Verify}(\text{vk}, \rho_i, \sigma_y^*)$
    if $y^* \neq \bot \wedge y^* \neq f(x)$ **then**
      $\delta_j := 1$
    else
      $\delta_j := 0$
    end if
  end for
  if $\exists j$ such that $\delta_j = 1$ **then**
    return 1
  else

return 0
end if

In the non-adaptive case the adversaries A's advantage is defined as

$$\text{Adv}_A^{\text{Verify}}(VC, f, \lambda) = \Pr\left[\text{EXP}_A^{\text{Verify}}[VC, f, \lambda] = 1\right].$$

So in practice this type of adversary is acceptable if a client aborts the protocol once it detects an incorrect result.

An adaptive adversaries A's advantage is defined as

$$\text{Adv}_A^{\text{AdaptVerify}}(VC, f, \lambda) = \Pr\left[\text{EXP}_A^{\text{AdaptVerify}}[VC, f, \lambda] = 1\right].$$

From this the security definition for verifiable computing schemes follows.

Definition 2.5 (Security)   A verifiable computing scheme VC is (weakly) secure if

$$\text{Adv}_A^{\text{Verify}}(VC, f, \lambda) \leq \text{negl}(\lambda)$$

and adaptively secure if

$$\text{Adv}_A^{\text{AdaptVerify}}(VC, f, \lambda) \leq \text{negl}(\lambda).$$

2.2.2 Privacy

Verifiable computing can guarantee the integrity of a computation. Another desirable property is to protect the secrecy of the client's inputs towards the server and when using a publicly verifiable scheme also towards the verifiers. To formally define *input privacy w.r.t the server* we define the following experiment. We use the oracle $O^{\text{ProbGen(sk},x)}$ which calls $\text{ProbGen(sk}, x)$ to obtain (σ_x, ρ_x) and only returns the public part σ_x.

Experiment $\text{EXP}_A^{\text{PrivacyServer}}[VC, f, \lambda]$
   $(\text{sk}, \text{vk}, \text{ek}) \leftarrow \text{KeyGen}(f, 1^\lambda)$
   $(x_0, x_1) \leftarrow A^{O^{\text{ProbGen(sk},\cdot)}}(\text{ek})$
   $(\sigma_0, \rho_0) \leftarrow \text{ProbGen(sk}, x_0)$
   $(\sigma_1, \rho_1) \leftarrow \text{ProbGen(sk}, x_1)$
   $b \xleftarrow{\$} \{0, 1\}$
   $b^* \leftarrow A^{O^{\text{ProbGen(sk},\cdot)}}(\text{ek}, x_0, x_1, \sigma_b)$
   if $b^* = b$ **then**
      return 1
   else
      return 0
   end if

In this experiment, the adversary first receives the public evaluation key for the scheme. Then, it selects two inputs x_0, x_1 and is given the encoding of one of the two inputs chosen at random. The adversary then must determine which input has been encoded. Note that during this process the adversary is allowed to request the encoding of any input of its choice. We define an adversaries A's advantage as

$$\mathsf{Adv}_A^{\mathsf{PrivacyServer}}(VC, f, \lambda) = \left| \Pr\left[\mathbf{EXP}_A^{\mathsf{PrivacyServer}}[VC, f, \lambda] = 1 \right] - 1/2 \right|.$$

Definition 2.6 (Input Privacy w.r.t. the Server) A verifiable computing scheme VC provides input privacy if

$$\mathsf{Adv}_A^{\mathsf{PrivacyServer}}(VC, f, \lambda) \leq \mathsf{negl}(\lambda).$$

Besides *input privacy* a verifiable computing scheme can also provide privacy with respect to the data output. This so called *output privacy* can be defined by an analogous experiment and is omitted here.

If we have a publicly verifiable computing scheme a third party verifier might try to learn about the input data from the publicly available verification data. To formally define *input privacy w.r.t a third party verifier* we define the following experiment.

Experiment $\mathbf{EXP}_A^{\mathsf{PrivacyVerifier}}[VC, f, \lambda]$
 $(\mathsf{sk}, \mathsf{vk}, \mathsf{ek}) \leftarrow \mathsf{KeyGen}(f, 1^\lambda)$
 $(x_0, x_1) \leftarrow A^{O^{\mathsf{ProbGen}(\mathsf{sk}, \cdot)}}(\mathsf{vk})$
 $(\sigma_0, \rho_0) \leftarrow \mathsf{ProbGen}(\mathsf{sk}, x_0)$
 $(\sigma_1, \rho_1) \leftarrow \mathsf{ProbGen}(\mathsf{sk}, x_1)$
 $b \xleftarrow{\$} \{0, 1\}$
 $b^* \leftarrow A^{O^{\mathsf{ProbGen}(\mathsf{sk}, \cdot)}}(\mathsf{vk}, x_0, x_1, \sigma_b, \rho_b)$
 if $b^* = b$ **then**
  return 1
 else
  return 0
 end if

In this experiment, the adversary first receives the public verification key for the scheme. Then, it selects two inputs x_0, x_1 and is given the encoding of one of the two inputs chosen at random. The adversary then must determine which input has been encoded. Note that during this process the adversary is allowed to request the encoding of any input of its choice.

We define an adversaries A's advantage as

$$\mathsf{Adv}_A^{\mathsf{PrivacyVerifier}}(VC, f, \lambda) = \left| \Pr\left[\mathbf{EXP}_A^{\mathsf{PrivacyVerifier}}[VC, f, \lambda] = 1 \right] - 1/2 \right|.$$

Definition 2.7 (Input Privacy w.r.t. the Verifier)   A verifiable computing scheme
VC provides input privacy if

$$\mathsf{Adv}_A^{\mathsf{PrivacyVerifier}}(VC, f, \lambda) \leq \mathsf{negl}(\lambda).$$

2.2.3   Efficiency

Finally we are interested in using verifiable computing schemes by means of
delegating computations. For this we want the work performed by the client and
the verifier to be less than computing the function on their own.

Definition 2.8 (Efficiency)   A verifiable computing scheme provides *efficiency* if
for any x and any σ_y, the time required for $\mathsf{KeyGen}(1^\lambda, f)$ plus the time required for
$\mathsf{ProbGen}(\mathsf{sk}, x)$ plus the time required for $\mathsf{Verify}(\mathsf{vk}, \rho_x, \sigma_y)$ is $o(T)$, where T is the
time required to compute $f(x)$.

A slightly relaxed definition is the following.

Definition 2.9 (Amortized Efficiency)   A verifiable computing scheme provides
amortized efficiency if it permits efficient verification. This implies that for any x
and any σ_y, the time required for $\mathsf{Verify}(\mathsf{vk}, \rho_x, \sigma_y)$ is $o(T)$, where T is the time
required to compute $f(x)$.

Note that in literature amortized efficiency has been defined ambiguously. We use
here a broad version that ensures that the minimal requirements for outsourceability
are met.

Intuitively the difference between efficiency and amortized efficiency is the cost
of the preprocessing phase. Efficient verifiable computing schemes allow a verifier
to verify the correctness of a computation more efficiently than performing the
computation by itself, including the preprocessing phase performed by the client.
Some verifiable computing schemes have an expensive preprocessing phase, but still
provide an efficient verification phase. Since the preprocessing phase only has to be
performed once and might not be time critical in many applications, we classify
them as verifiable computing schemes providing amortized efficiency.

One aspect that also impacts the practicality of all verifiable computing schemes
is the server's overhead to evaluate a computation using $\mathsf{Compute}$ versus natively
executing it. Note that this does not affect the computation complexity for the client
or the verifier. So far in literature this aspect has not been rigorously covered and is
therefore not considered in our efficiency analysis.

References

1. S. Benabbas, R. Gennaro, Y. Vahlis, Verifiable delegation of computation over large datasets, in
   Advances in Cryptology - CRYPTO 2011 - 31st Annual Cryptology Conference, Proceedings,
   Santa Barbara, CA, 14–18 August 2011, pp. 111–131

2. R. Gennaro, C. Gentry, B. Parno, Non-interactive verifiable computing: outsourcing computation to untrusted workers, in *Advances in Cryptology - CRYPTO 2010, 30th Annual Cryptology Conference, Proceedings*, Santa Barbara, CA, 15–19 August 2010, pp. 465–482
3. B. Parno, M. Raykova, V. Vaikuntanathan, How to delegate and verify in public: verifiable computation from attribute-based encryption, in *Theory of Cryptography - 9th Theory of Cryptography Conference, TCC 2012, Proceedings*, Taormina, Sicily, 19–21 March 2012, pp. 422–439

Chapter 3
Proof and Argument Based Verifiable Computing

Abstract In this chapter the state of the art with respect to proof based verifiable computing schemes is presented. In this setting a prover wants to convince a verifier of the correctness of a computed result. The first proof based solutions that achieve this were interactive proof systems. Depending on the computation power of the prover we distinguish here between proof based and argument based approaches. While all proof based schemes are interactive protocols, the argument based solutions were further improved, such that also non-interactive solutions are available. In this chapter, we first provide an introduction presenting the setting and the notions, i.e. quadratic span program (QSP), quadratic arithmetic program (QAP), and succinct non-interactive arguments of knowledge (SNARKs). Then, we present the interactive proof based solutions, i.e. "Verifiable Computation with Massively Parallel Interactive Proofs" by Thaler et al. and "Allspice" by Vu et al., and the argument based approaches, i.e. "Pepper" by Setty et al., "Ginger" by Setty et al., "Zaatar" by Setty et al., "Pantry" by Braun et al., and "River" by Xu et al. Afterwards, we present the definitions and solutions for the non-interactive argument based verifiable computing schemes, i.e. "Pinocchio" by Parno et al., "Geppetto" by Costello et al., "SNARKs for C" by Ben-Sasson et al., "Succinct Non-interactive Zero Knowledge for a von Neumann Architecture" by Ben-Sasson et al., "Buffet" by Wahby et al., "ADSNARK" by Backes et al., and "Block Programs: Improving Efficiency of Verifiable Computation for Circuits with Repeated Substructures" by Xu et al.

3.1 Introduction to Proof and Argument Based Approaches

In the setting of proof based verifiable computing a *(super-)polynomial-time prover* wants to convince a *computationally bounded verifier* of the validity of some statement in an **NP**-language. In the context of verifiable computing the prover is the server performing a given computation and the statement represents the correctness of the computed result. To achieve this goal, rather theoretical tools, such as *interactive proof systems* [16] and *probabilistically checkable proofs (PCP)* [1, 2] were used. While the application to verifiable computing scenarios have already been mentioned in very early works, the solutions from these theoretical tools were not suitable for any practical application. Later work relaxed these

© The Author(s) 2017   13
D. Demirel et al., *Privately and Publicly Verifiable Computing Techniques*,
SpringerBriefs in Computer Science, DOI 10.1007/978-3-319-53798-6_3

potentially super-polynomial provers to polynomially bounded provers to obtain (typically more efficient) *argument systems*. Furthermore, this line of work has been improved by both theoretical refinements and suitable implementations. To verify the correctness of an outsourced program or function by using proof (or argument) based systems, the program or function has to be encoded. A suitable encoding is to write it as a circuit or to express it as a set of arithmetic constraints, i.e. polynomials which simultaneously evaluate to 0 iff the circuit is evaluated correctly. For the latter case, Gennaro et al. [15] introduced two new notions, one called quadratic span program (QSP) for boolean circuits and another one for arithmetic circuits called quadratic arithmetic program (QAP). These constructions have been developed specifically for the verifiable computing use case. The basic idea is to write a circuit as a set of degree-2 constraints over some large finite field. As later shown in [9] the QAP approach in [15] implicitly uses a PCP structure. While for the proof based solutions only interactive protocols are available, there are argument based approaches available that are non-interactive. To achieve this Gennaro et al. introduced in [15] a construct for succinct non-interactive arguments of knowledge (SNARKs) using QSPs and QAPs. In the following the existing interactive proof based solutions and interactive and non-interactive argument based solutions are presented. We refer the interested reader to a recent article by Walfish and Blumberg [33] that provides a good overview of proof based approaches to verifiable computations as well as the available tools basing on different approaches.

3.2 Interactive Proof Based Approaches

In the following interactive proof systems are considered. First, the required concepts are defined followed by the different solutions. Let $L \subseteq \{0, 1\}^*$ be an **NP**-language, then interactive proof systems are defined as follows.

Definition 3.1 (Interactive Proof System (IPS)) An *interactive proof system for a language L* is an interactive protocol between an unrestricted prover P and a PPT verifier V, such that the following conditions hold:

Completeness. $\forall x \in L : \Pr[\, (P, V)(x) = 1\,] = 1$,
Soundness. $\forall x \notin L \, \forall P^* : \Pr[\, (P^*, V)(x) = 1\,] \leq \frac{1}{2}$,

where we use $(P, V)(x) = 1$ to denote that V accepts the interaction with P on common input x.

Let $R \subseteq \{0, 1\}^* \times \{0, 1\}^*$ be a polynomial-time (witness) relation, i.e. a relation such that membership of (x, w) in R can be decided in polynomial time in $|x|$. In the following, for an **NP**-language L we may explicitly index it with its witness relation and write L_R where $L_R = \{x \mid \exists w : (x, w) \in R\}$. Now, we can define the concept of proofs of knowledge which defines IPs with a stronger notion of soundness.

While the traditional definition of IPS does not put any restrictions on the prover P and in particular allows P to run in super-polynomial time, this is clearly not meaningful for the application to verifiable computing. In [17] Goldwasser et al.

present efficient interactive proofs with polynomial provers for any function representable as a log-space uniform circuit that has communication complexity being the depth of the circuit. For refinements of this, prototypical implementations are available at [18].

Several subsequent work build on verifiable computing frameworks based on such IPS with polynomial provers. These are more efficient than the other approaches discussed later in this chapter, but their expressibility in terms of the computations provided is rather limited. In [25] constant round proofs suitable for outsourcing of computations are presented.

In the following we present the two existing interactive proof based approaches. Both are secure against a strong adversary, provide public verifiability, but do not address privacy.

3.2.1 Verifiable Computation with Massively Parallel Interactive Proofs

In [30] Thaler et al. proposed the first verifiable computing protocol with a reasonable server's overhead. The approach is to use parallel processing, i.e. running parts of the protocol in parallel using a GPU, to speed up the evaluation process. Their protocol supports arithmetic circuits of polylogarithmic depth. In [29] the authors define the notion of *regularity* of a function. This contains for instance to what extent output bits depend on input bits and therefore to what extent the computation of the function can be parallelised. For circuits that are *regular* in this sense the server's overhead is just a factor of approximately 10. It follows that for those circuits this verifiable computing scheme even provides efficiency. In addition, this construction does not depend on any cryptographic assumption.

3.2.2 Allspice: A Hybrid Architecture for Interactive Verifiable Computation

In [31] Vu et al. generalized [30] to functions beyond arithmetic circuits. They build a system called *Allspice* that also supports comparisons and inequality checks. This allows to verify computations expressed as straight-line programs (i.e. programs that do neither branch nor loop). Furthermore, their schemes improves with respect to the server's overhead for *non-regular* functions. On the other hand, they require a computationally more expensive setup phase. Therefore, this scheme only achieves amortized efficiency. For Allspice a prototypical implementation is available at [18].

3.3   Interactive Argument Based Approaches

While in the IPS setting the soundness guarantees are unconditional, i.e. hold
with respect to an all-powerful prover, one may reduce the soundness guarantees
to computational soundness, i.e. computationally bounded provers. The resulting
systems are no longer denoted as proofs, but as arguments.

In the following probabilistically checkable proofs (PCPs) are defined, because
they are (implicitly) used in all approaches [15]. These proofs can be verified by
a randomized algorithm using a bounded number of random coins and inspecting a
bounded number of bits in the proof.

Definition 3.2 (Probabilistically Checkable Proof (PCP))   A probabilistically
checkable proof (PCP) for a language $L \in \mathbf{PCP}(r(n), q(n))$ is a string π such that
there exists a PPT algorithm V (the verifier) that, on input $a \in \{0, 1\}^n$ uses $O(r(n))$
random coins and inspects $O(q(n))$ locations in π, after which it outputs 1 (accept)
or 0 (reject) such that:

Completeness.   If $a \in L$, then there exists a π such that $\Pr[V^{\pi}(a) = 1] = 1$.

Soundness.   If $a \notin L$, then for all π^* it holds that $\Pr\left[V^{\pi^*}(a) = 1\right] < 1/2$.

Here V^{π} denotes that V has oracle access to the string π.

While asymptotically short PCPs [4, 5] are interesting in theory, for their
application to verifiable computing the length of the PCP (needed to be retrieved
by the verifier) is still longer than the execution trace of any function. Thus, this
does not yield solutions that provide verification that is more efficient than the local
evaluation of the function.

The idea of computationally sound proofs dates back to Kilian [22], who
proposed to combine PCPs with linear commitments having local openings (such
as those obtained from collision resistant hash functions and generally known as
Merkle Trees). The prover commits to a PCP string π and sends the commitment to
π to the verifier. Then, the verifier can ask the prover to open the commitment on
various positions (determined by the random coins of the PCP verifier). So, one
obtains four-move argument systems. Later Micali has shown in [23] how this
approach can be turned into a one-move scheme secure in the random oracle model
by applying the Fiat-Shamir heuristic [13]. The basic idea is simply to let the prover
compute the random coins (of the PCP verifier) by computing them from the output
of a random oracle having the commitment to π as input.

Another direction within interactive arguments is the use of what is called linear
PCPs [21]. The PCP string is implicitly represented as a linear function and the
verifier uses additively homomorphic encryption to commit to a function of this
form (cf. [21, 26]). Although, this comes at the cost of an expensive preprocessing
stage this can be amortized over a batch of verifications of the same function over
different inputs.

In the following the different interactive argument based approaches are presented. All schemes achieve only amortized efficiency and do not address privacy. Their security has not been rigorously analysed yet.

3.3.1 Pepper: Making Argument Systems for Outsourced Computation Practical (Sometimes)

In 2012 Setty et al. [26] present an interactive argument system named *Pepper*. Here functions are not represented as circuits but as arithmetic constraints. These are algebraic equations that hold simultaneously if and only if function f is evaluated correctly. Pepper only supports a very limited class of functions and only achieves amortized efficiency, while having a setup phase, whose computational cost is proportional to $O(|f|)$. Also for this solution a prototypical implementation is available (see [18]).

3.3.2 Ginger: Taking Proof-Based Verified Computation a Few Steps Closer to Practicality

Setty et al. further improved on [26] in a system called Ginger [27] that supports a larger class of computations such as inequality tests, floating point arithmetics, and conditional branching. A prototypical implementation for this solution is available (see [18]).

3.3.3 Zaatar: Resolving the Conflict Between Generality and Plausibility in Verified Computation

Setty et al. [28] developed an improvement over Pepper and Ginger called Zaatar. It uses the algebraic representation of computations introduced as QAPs in [15] and removes the restrictions of the previous schemes, i.e. this new PCP supports a richer class of functions. Furthermore, it is also shown that Zaatar improves on the efficiency of Ginger. For this solution a prototypical implementation is provided (see [18]).

3.3.4 Pantry: Verifying Computations with State

Braun et al. introduced within their construction Pantry [10] an expansion of Zaatar which allows verification of stateful computations. For a prototypical implementation see [18].

3.3.5 River: Verifiable Computation with Reduced Informational Costs and Computational Costs

In [34] Xu et al. present a QAP based verifiable computing system named River. Compared to Zaatar, River reduces the client's computational costs while only marginally increasing the server's overhead. This scheme supports arithmetic circuits and achieves amortized efficiency.

3.4 Non-interactive Argument Based Approaches

All proof based schemes presented so far are interactive protocols. In order to provide a non-interactive solution, Gennaro et al. show in [15] how to construct succinct non-interactive arguments of knowledge (SNARKs) using QSPs and QAPs. Like for all QAP based schemes even though this primitive is secure against the adaptive adversary it is an open task to prove that also the verifiable computing techniques using this primitive provide the same level of security. Furthermore, all these schemes are only secure under an assumption that is *non-falsifiable*, i.e. they cannot be verified nor denied.

Definition 3.3 (Succinct Non-interactive Argument (SNARG)) [8]) A SNARG for the relation $R \subset R_U$, where R_U is the "universal relation", is a triple of the following probabilistic, polynomial-time algorithms:

- $\mathsf{Gen}_V(1^\lambda) \rightarrow (\mathsf{vgrs}, \mathsf{priv})$. Takes the security parameter λ as input and outputs a verifier-generated reference string vgrs and corresponding private verification coins priv.
- $\mathsf{P}(y, w, \mathsf{vgrs}) \rightarrow \pi$. Takes a statement $y = (M, x, t)$, a witness w, and the reference string vgrs and outputs a proof π.
- $\mathsf{V}(\mathsf{priv}, y, \pi) \rightarrow \{0, 1\}$ verifies the validity of π for y using the private verification coins priv and returns '1' if the input is correct and '0' otherwise.

These algorithms have to satisfy the following conditions.

Completeness. For any $(y, w) \in R$

$$\Pr\big[\mathsf{V}(\mathsf{priv}, y, \pi) = 1 \mid (\mathsf{vgrs}, \mathsf{priv}) \leftarrow \mathsf{Gen}_V(1^\lambda), \pi \leftarrow \mathsf{P}(y, w, \mathsf{vgrs})\big] = 1$$

Succinctness. The length of π that $\mathsf{P}(y, w, \mathsf{vgrs})$ outputs as well as the running time of $\mathsf{V}(\mathsf{priv}, y, \pi)$ is bounded by

$$p(\lambda + |y|) = p(\lambda + |M| + |x| + log(t)),$$

where p is a universal polynomial that does not depend on R.

Adaptive Soundness.   For all poly-size prover P^* and large enough $\lambda \in \mathbb{N}$

$$\Pr\big[V(\mathsf{priv}, y, \pi) = 1 \mid (\mathsf{vgrs}, \mathsf{priv}) \leftarrow \mathsf{Gen}_V(1^\lambda), (y, \pi) \leftarrow P^*(\mathsf{vgrs}), y \notin L_R\big]$$
$$\leq \mathsf{negl}(\lambda).$$

For our purpose we need an even stronger definition.

Definition 3.4 (SNARG of Knowledge (SNARK) [8]) A SNARK is a SNARG (Gen_V, P, V) where soundness is replaced by the following stronger condition.

Adaptive Proof of Knowledge.   For any poly-size prover P^* there exists a poly-size extractor E_{P^*} such that for all large enough $\lambda \in \mathbb{N}$ and all auxiliary inputs $z \in \{0, 1\}^{\mathsf{poly}(\lambda)}$

$$\Pr\left[\begin{array}{c} (\mathsf{vgrs}, \mathsf{priv}) \leftarrow \mathsf{Gen}_V(1^\lambda) \\ (y, \pi) \leftarrow P^*(z, \mathsf{vgrs}) \quad \wedge \quad \begin{array}{c} (y, w) \leftarrow E_{P^*}(z, \mathsf{vgrs}) \\ w \notin R(y) \end{array} \\ V(\mathsf{priv}, y, \pi) = 1 \end{array}\right] \leq \mathsf{negl}(\lambda).$$

There also exists the notion of *O-SNARKs* (see [14]), which formalizes SNARKs in the presence of oracles. In the following we present the different non-interactive argument based approaches. All of them achieve only amortized efficiency, but provide public verifiability, security against adaptive adversaries, and input privacy.

3.4.1   Pinocchio: Nearly Practical Verifiable Computation

In [24] Parno et al. developed a system named Pinocchio that supports arithmetic circuits (that are turned into QAPs). Pinocchio offers public verifiability, but no privacy towards the server. However there exists a variant of Pinocchio that offers input privacy towards the (public) verifier. Due to its preprocessing phase that runs in time proportional to a one time execution of function f it only achieves amortized efficiency. For this solution a prototypical implementation is available at [19]. Its security is based on the q-PDH assumption (see Assumption A.6), the q-SDH assumption (see Assumption A.6), and the q-PKE assumption (see Assumption A.7).

3.4.2   Geppetto: Versatile Verifiable Computation

Costello et al. generalized the QAPs to MultiQAPs and use them to build a verifiable computing system called Geppetto [12]. They show how to reduce the server's overhead by decomposing circuits into a collection of subcircuits. Geppetto offers public verifiability, but no privacy towards the server, while allowing for input privacy towards the (public) verifier. Due to its preprocessing phase that runs in time proportional to a one time execution of function f it only achieves amortized efficiency. Also for this solution a prototypical implementations is available (see [19]).

3.4.3   SNARKs for C: Verifying Program Executions Succinctly and in Zero Knowledge

In [6] Ben-Sasson et al. present a system that can verify all operations in programming language C albeit at an increased server's overhead compared to [24]. It is also based on QAPs. This system also offers public verifiability without privacy towards the server while guaranteeing input privacy towards the verifier. It achieves amortized efficiency and a prototypical implementation for this solution is available (see [20]).

3.4.4   Succinct Non-interactive Zero Knowledge for a von Neumann Architecture

In [7] Ben-Sasson et al. present a new QAP based SNARK for arithmetic circuits that allows for more efficient verification and proof generation compared to [24] and [6]. They also provide a universal circuit generator further broadening the class of admitted programs. Furthermore, they show that their approach provides zero-knowledge, i.e. if the statement is true, no cheating verifier learns anything other than this. Their construction is publicly verifiable. However, they provide only input-output privacy with respect to the verifier, but not towards the server performing the operation. For this solution a prototypical implementation is available (see [20]).

3.4.5   Buffet: Efficient RAM and Control Flow in Verifiable Outsourced Computation

In [32] Wahby et al. improve upon the functionality by supporting programs with general loops. It can be made publicly verifiable. For this solution a prototypical implementation is available (see [18]).

3.4.6   ADSNARK: Nearly Practical and Privacy-Preserving Proofs on Authenticated Data

Backes et al. present ADSNARK [3] a non-interactive proof system for straight-line computations on authenticated data. Following the generic construction presented in [11] one obtains a verifiable computing system. ADSNARK includes both a publicly verifiable and a more efficient privately verifiable proof. It achieves amortized efficiency and input privacy with respect to the verifier. However, note that this does not prevent the server performing the computation from seeing the input values. For this solution a prototypical implementation is available (see [20]).

3.4.7 Block Programs: Improving Efficiency of Verifiable Computation for Circuits with Repeated Substructures

Xu et al. showed in [35] a new and more efficient way to handle loops in a program. This improvement can be used together with all other approaches listed here that support loops.

References

1. S. Arora, S. Safra, Probabilistic checking of proofs: a new characterization of NP. J. ACM **45**, 70–122 (1998)
2. L. Babai, L. Fortnow, L.A. Levin, M. Szegedy, Checking computations in polylogarithmic time, in *Proceedings of the 23rd Annual ACM Symposium on Theory of Computing* (1991), pp. 21–31
3. M. Backes, M. Barbosa, D. Fiore, R.M. Reischuk, ADSNARK: nearly practical and privacy-preserving proofs on authenticated data, in *2015 IEEE Symposium on Security and Privacy, SP 2015*, San Jose, CA, 17–21 May 2015, pp. 271–286
4. E. Ben-Sasson, O. Goldreich, P. Harsha, M. Sudan, S.P. Vadhan, Short PCPs verifiable in polylogarithmic time, in *20th Annual IEEE Conference on Computational Complexity (CCC 2005)* (2005), pp. 120–134
5. E. Ben-Sasson, O. Goldreich, P. Harsha, M. Sudan, S.P. Vadhan, Robust PCPs of proximity, shorter PCPs, and applications to coding. SIAM J. Comput. **36**, 889–974 (2006)
6. E. Ben-Sasson, A. Chiesa, D. Genkin, E. Tromer, M. Virza, SNARKs for C: verifying program executions succinctly and in zero knowledge, in *Advances in Cryptology - CRYPTO 2013 - 33rd Annual Cryptology Conference, Proceedings, Part II*, Santa Barbara, CA, 18–22 August 2013, pp. 90–108
7. E. Ben-Sasson, A. Chiesa, E. Tromer, M. Virza, Succinct non-interactive zero knowledge for a von Neumann architecture, in *Proceedings of the 23rd USENIX Security Symposium*, San Diego, CA, 20–22 August 2014, pp. 781–796
8. N. Bitansky, R. Canetti, A. Chiesa, E. Tromer, From extractable collision resistance to succinct non-interactive arguments of knowledge, and back again, in *Innovations in Theoretical Computer Science 2012*, Cambridge, MA, 8–10 January 2012, pp. 326–349
9. N. Bitansky, A. Chiesa, Y. Ishai, R. Ostrovsky, O. Paneth, Succinct non-interactive arguments via linear interactive proofs, in *TCC* (2013), pp. 315–333
10. B. Braun, A.J. Feldman, Z. Ren, S.T.V. Setty, A.J. Blumberg, M. Walfish, Verifying computations with state, in *ACM SIGOPS 24th Symposium on Operating Systems Principles, SOSP '13*, Farmington, PA, 3–6 November 2013, pp. 341–357
11. R. Canetti, B. Riva, G.N. Rothblum, Two protocols for delegation of computation, in *Information Theoretic Security - 6th International Conference, ICITS 2012, Proceedings*, Montreal, QC, 15–17 August 2012, pp. 37–61
12. C. Costello, C. Fournet, J. Howell, M. Kohlweiss, B. Kreuter, M. Naehrig, B. Parno, S. Zahur, Geppetto: versatile verifiable computation, in *2015 IEEE Symposium on Security and Privacy, SP 2015*, San Jose, CA, 17–21 May 2015, pp. 253–270
13. A. Fiat, A. Shamir, How to prove yourself: practical solutions to identification and signature problems, in *CRYPTO* (1986), pp. 186–194
14. D. Fiore, A. Nitulescu, On the (in)security of SNARKs in the presence of oracles, in *Theory of Cryptography - 14th International Conference, TCC 2016-B, Proceedings, Part I*, Beijing, 31 October–3 November 2016, pp. 108–138
15. R. Gennaro, C. Gentry, B. Parno, M. Raykova, Quadratic span programs and succinct NIZKs without PCPs, in *Advances in Cryptology - EUROCRYPT 2013, 32nd Annual International*

Conference on the Theory and Applications of Cryptographic Techniques, Proceedings, Athens, 26–30 May 2013, pp. 626–645

16. S. Goldwasser, S. Micali, C. Rackoff, The knowledge complexity of interactive proof systems. SIAM J. Comput. **18**, 186–208 (1989)

17. S. Goldwasser, Y.T. Kalai, G.N. Rothblum, Delegating computation: interactive proofs for muggles, in *Proceedings of the 40th Annual ACM Symposium on Theory of Computing,* Victoria, BC, 17–20 May 2008, pp. 113–122

18. http://cs.utexas.edu/pepper. Retrieved 18 Apr 2016

19. http://research.microsoft.com/verifcomp/. Retrieved 18 Apr 2016

20. https://github.com/scipr-lab/libsnark. Retrieved 18 Apr 2016

21. Y. Ishai, E. Kushilevitz, R. Ostrovsky, Efficient arguments without short PCPs, in *22nd Annual IEEE Conference on Computational Complexity (CCC 2007),* San Diego, CA, 13–16 June 2007, pp. 278–291

22. J. Kilian, A note on efficient zero-knowledge proofs and arguments (extended abstract), in *Proceedings of the 24th Annual ACM Symposium on Theory of Computing,* Victoria, BC, 4–6 May 1992, pp. 723–732

23. S. Micali, Computationally sound proofs. SIAM J. Comput. **30**, 1253–1298 (2000)

24. B. Parno, J. Howell, C. Gentry, M. Raykova, Pinocchio: nearly practical verifiable computation, in *2013 IEEE Symposium on Security and Privacy, SP 2013,* Berkeley, CA, 19–22 May 2013, pp. 238–252

25. O. Reingold, G.N. Rothblum, R.D. Rothblum, Constant-round interactive proofs for delegating computation, in *Proceedings of the 48th Annual ACM SIGACT Symposium on Theory of Computing, STOC 2016,* Cambridge, MA, 18–21 June 2016, pp. 49–62

26. S.T.V. Setty, R. McPherson, A.J. Blumberg, M. Walfish, Making argument systems for outsourced computation practical (sometimes), in *19th Annual Network and Distributed System Security Symposium, NDSS 2012,* San Diego, CA, 5–8 February 2012

27. S.T.V. Setty, V. Vu, N. Panpalia, B. Braun, A.J. Blumberg, M. Walfish, Taking proof-based verified computation a few steps closer to practicality, in *Proceedings of the 21th USENIX Security Symposium,* Bellevue, WA, 8–10 August 2012, pp. 253–268

28. S.T.V. Setty, B. Braun, V. Vu, A.J. Blumberg, B. Parno, M. Walfish, Resolving the conflict between generality and plausibility in verified computation, in *Eighth Eurosys Conference 2013, EuroSys '13,* Prague, 14–17 April 2013, pp. 71–84

29. J. Thaler, Time-optimal interactive proofs for circuit evaluation, in *Advances in Cryptology - CRYPTO 2013 - 33rd Annual Cryptology Conference, Proceedings, Part II,* Santa Barbara, CA, 18–22 August 2013, pp. 71–89

30. J. Thaler, M. Roberts, M. Mitzenmacher, H. Pfister, Verifiable computation with massively parallel interactive proofs, in *4th USENIX Workshop on Hot Topics in Cloud Computing, HotCloud'12,* Boston, MA, 12–13 June 2012

31. V. Vu, S.T.V. Setty, A.J. Blumberg, M. Walfish, A hybrid architecture for interactive verifiable computation, in *2013 IEEE Symposium on Security and Privacy, SP 2013,* Berkeley, CA, 19–22 May 2013, pp. 223–237

32. R.S. Wahby, S.T.V. Setty, Z. Ren, A.J. Blumberg, M. Walfish, Efficient RAM and control flow in verifiable outsourced computation, in *22nd Annual Network and Distributed System Security Symposium, NDSS 2015,* San Diego, CA, 8–11 February 2015

33. M. Walfish, A.J. Blumberg, Verifying computations without reexecuting them. Commun. ACM **58**, 74–84 (2015)

34. G. Xu, G.T. Amariucai, Y. Guan, Verifiable computation with reduced informational costs and computational costs, in *Computer Security - ESORICS 2014 - 19th European Symposium on Research in Computer Security, Proceedings, Part I,* Wroclaw, 7–11 September 2014, pp. 292–309

35. G. Xu, G.T. Amariucai, Y. Guan, Block programs: improving efficiency of verifiable computation for circuits with repeated substructures, in *Proceedings of the 10th ACM Symposium on Information, Computer and Communications Security, ASIA CCS '15,* Singapore, 14–17 April 2015, pp. 405–416

Chapter 4
Verifiable Computing from Fully Homomorphic Encryption

Abstract In this chapter we discuss approaches to verifiable computing that use fully homomorphic encryption (FHE) as a building block. First, we define homomorphic encryption and fully homomorphic encryption. Then, we describe the verifiable computing schemes using this primitive, i.e. "Non-Interactive Verifiable Computing: Outsourcing Computation to Untrusted Workers" by Gennaro et al., "Improved Delegation of Computation Using Fully Homomorphic Encryption" by Chung et al., and "Efficient Non-Interactive Verifiable Outsourced Computation for Arbitrary Functions" by Chen et al. Note that using these solutions the client encrypts the data before it outsources it to the server. Thus, these solutions achieve input privacy. In addition, only the client can decrypt the result, which is why also output privacy is assured. However, on the downside all fully homomorphic encryption based schemes are only privately verifiable. Furthermore, all solutions are only secure against weak adversaries and providing efficient FHE schemes is still an open research challenge.

4.1 Definitions for Fully Homomorphic Encryption

Definition 4.1 (Homomorphic Encryption (HE) Scheme [3]) A *homomorphic encryption scheme* is a tuple of the following probabilistic, polynomial-time algorithms:

KeyGen(1^λ) : This algorithm takes a security parameter λ as input and outputs a public key pk and a secret key sk. The public key pk implicitly defines a message space M, and a ciphertext space C.

Encrypt(pk, m) : The encryption algorithm takes a public key pk and message $m \in M$ as input and outputs a ciphertext c.

Decrypt(sk, c) : The decryption takes a secret key sk and a ciphertext c as input and outputs a message $m \in M \cup \perp$.

Eval(pk, f, $\mathbf{c}_i$) : The evaluation algorithm takes a public key pk, a description of a function f, and a vector of ciphertexts $\mathbf{c}_i$ as input and outputs a new ciphertext c.

© The Author(s) 2017

D. Demirel et al., *Privately and Publicly Verifiable Computing Techniques*,
SpringerBriefs in Computer Science, DOI 10.1007/978-3-319-53798-6_4

A *homomorphic encryption scheme* is homomorphic for a class F of functions, if

$$\forall f \in F, \{m_i\} \subset M$$

$$\Pr[\text{Decrypt}(\text{sk}, \text{Eval}(\text{pk}, f, \{\text{Encrypt}(\text{pk}, m_i)\})) = f(m_1, \ldots, m_n)] = 1.$$

Besides the above property of evaluating correctness (which may also allow a negligible evaluation error), one requires the usual correctness property of an encryption scheme as well as at least IND-CPA security.

Now, informally, a HE scheme is called fully homomorphic (is a FHE scheme) if the class F of functions represents the class of all circuits. We stress that one requires an additional compactness property, which basically means that the ciphertext output by the Eval algorithm does only depend on the security parameter (and not on the function). This rules out trivial constructions of FHE, e.g. ones where the Eval algorithm simply applies the identity function (or does nothing) and the Decrypt algorithm evaluates the function on the decrypted ciphertext(s) and then returns the result. We do not require a formal treatment of properties of FHE here and refer the reader to [1] for an overview.

4.2 Verifiable Computing Schemes Based on FHE

4.2.1 Non-interactive Verifiable Computing: Outsourcing Computation to Untrusted Workers

The verifiable computing scheme presented in [3] by Gennaro et al. achieves verifiability by combining Yao's garbled circuits [5, 6] with FHE. This combination allows to reuse a garbled circuit multiple times while still preserving security. The idea is that during the setup, the client once generates a garbled version of a circuit representing a function f and sets the public key to the garbled circuit and the secret key to the secret random wire labels. If the client wants to outsource a computation of f on some input x, it generates a fresh key pair (sk, pk) of an FHE scheme, sets the public value σ_x to pk, sets the decoding value ρ_x to sk, and generates ciphertexts to all wire values of the binary expression of the input x. Then, the server can use the homomorphic property of the FHE scheme to evaluate the received garbled circuit and to send the encrypted output wires back to the client. The client can then decrypt and map the wires to the output $y = f(x)$.

Besides providing privacy, for this construction the authors were the first to formally introduce the notion of verifiable computation (see Sect. 2.1). This scheme provides amortized efficiency, but the server's overhead depends on the efficiency of the underlying FHE scheme making it not practically efficient today. It only offers security against a weak adversary as no verification queries are allowed.

4.2.2 Improved Delegation of Computation Using Fully Homomorphic Encryption

In [2] Chung et al. presented another way to verify the correctness of a result by using FHE. The underlying idea is to evaluate f on some random point r in the preprocessing, i.e. to generate $y_r = f(r)$, and store this data for verification. More precisely, in the online phase the server is asked to compute and return f evaluated on both x and r in a random order. Then, the client can check if y_r equals the value returned for the computation corresponding to r. If this result is correct, the server is assumed to behave honestly. However, with this naive approach firstly the soundness error is very large, i.e. $1/2$, and secondly the precomputed value y_r can only be used once. In order to overcome these issues, the client precomputes $f(r_1, \ldots, r_n)$ for some random r_i and target function f for large enough n (to make the soundness error small enough). In addition, it uses a FHE scheme to compute encryptions $\hat{x}_i$ of the inputs x_i and the encryptions $\hat{r}_i$ of the random values r_i. The server will then be asked to homomorphically evaluate $f(\hat{x}_1, \ldots, \hat{x}_n)$ and $f(\hat{r}_1, \ldots, \hat{r}_n)$. The client can decrypt both results and accepts the computation as correct if one of them matches its precomputed result. This scheme achieves amortized efficiency while the server's overhead depends on the underlying FHE scheme. This scheme offers only weak security.

In [4] a similar scheme is presented, that reduces the preprocessing stage. It offers weak security. However it should be noted that its security against an adaptive adversary has not been analysed yet.

References

1. F. Armknecht, C. Boyd, C. Carr, K. Gjøsteen, A. Jäschke, C.A. Reuter, M. Strand, *A Guide to Fully Homomorphic Encryption*. Cryptology ePrint Archive, Report 2015/1192 (2015), http://eprint.iacr.org/
2. K. Chung, Y.T. Kalai, S.P. Vadhan, Improved delegation of computation using fully homomorphic encryption, in *Advances in Cryptology - CRYPTO 2010, 30th Annual Cryptology Conference, Proceedings*, Santa Barbara, CA, 15–19 August 2010, pp. 483–501
3. R. Gennaro, C. Gentry, B. Parno, Non-interactive verifiable computing: outsourcing computation to untrusted workers, in *Advances in Cryptology - CRYPTO 2010, 30th Annual Cryptology Conference, Proceedings*, Santa Barbara, CA, 15–19 August 2010, pp. 465–482
4. C. Tang, Y. Chen, *Efficient Non-interactive Verifiable Outsourced Computation for Arbitrary Functions*. IACR Cryptology ePrint Archive (2014), p. 439
5. A.C. Yao, Protocols for secure computations (extended abstract), in *23rd Annual Symposium on Foundations of Computer Science*, Chicago, IL, 3–5 November 1982, pp. 160–164
6. A.C. Yao, How to generate and exchange secrets (extended abstract), in *27th Annual Symposium on Foundations of Computer Science*, Toronto, 27–29 October 1986, pp. 162–167

Chapter 5
Homomorphic Authenticators

Abstract Homomorphic authenticators allow to evaluate functions on *authenticated* data. There exist constructions both in the secret key setting in the form of *homomorphic message authentication codes (MACs)* and in the public key setting in the form of *homomorphic signatures*. These solutions can be used to respectively construct *privately* and *publicly verifiable computing schemes*. There are homomorphic MAC and signature schemes that are not known to allow verification faster than computing the function, e.g. Gennaro and Wichs (Fully homomorphic message authenticators, in *Advances in Cryptology - ASIACRYPT 2013 - 19th International Conference on the Theory and Application of Cryptology and Information Security, Proceedings, Part II*, Bengaluru, 1–5 December 2013, pp. 301–320) or Freeman (Improved security for linearly homomorphic signatures: a generic framework, in *Public Key Cryptography - PKC 2012 - 15th International Conference on Practice and Theory in Public Key Cryptography, Proceedings*, Darmstadt, 21–23 May 2012, pp. 697–714), and are therefore not considered in this chapter. In the following, first, we provide the definitions for schemes using homomorphic authenticators and their correctness and security. Then we present privately verifiable computing schemes using MACs, i.e. "Verifiable Delegation of Computation on Outsourced Data" by Backes et al., "Generalized Homomorphic MACs with Efficient Verification" by Zhang and Safavi-Naini, and "Efficiently Verifiable Computation on Encrypted Data" by Fiore et al. Afterwards, we present the publicly verifiable computing schemes using homomorphic signatures, i.e. "Programmable Hash Functions Go Private" by Catalano et al., "Homomorphic Signatures with Efficient Verification for Polynomial Functions" by Catalano et al., and "Algebraic (Trapdoor) One-Way Functions and their Applications" by Catalano et al. Finally, we present an approach by Lai et al., "Verifiable Computation on Outsourced Encrypted Data", showing how to combine signature based verifiable computing with homomorphic encryption assuring privacy of the data processed.

5.1 Definitions for Homomorphic Authenticators

First, we provide the definitions for homomorphic authenticators, their correctness, and their security. To do so we use multi-labels and multilabeled programs, which we will briefly explain here. The labels are used to tag a dataset.

© The Author(s) 2017   27
D. Demirel et al., *Privately and Publicly Verifiable Computing Techniques*,
SpringerBriefs in Computer Science, DOI 10.1007/978-3-319-53798-6_5

A multi-label $L = (\Delta, \tau)$ consists of a data set identifier Δ and an input identifier τ. Given some function $f : M^n \to M$ that takes n inputs the labels $\tau_1, \ldots \tau_n$ label the different input columns while Δ labels the set from which we take our data. This allows to both identify the dataset a server is supposed to work on and to restrict the server to this data. A labeled program $P = (f, \tau_1, \ldots, \tau_n)$ consists of a function $f : M^n \to M$ on n variables and each $\tau_i \in \{0, 1\}^*$ is the label of the ith input to f for $i = 1, \ldots, n$. A multi-labeled program P_Δ is a pair (P, Δ) where $P = (f, \tau_1, \ldots, \tau_n)$ is a labeled program and $\Delta \in \{0, 1\}^*$ is the data set identifier.

Definition 5.1 (Homomorphic Authenticators) A scheme based on *homomorphic authenticators* is a tuple of the following probabilistic, polynomial-time algorithms:

KeyGen($1^\lambda, L$) :   The key generation algorithm takes as input a security parameter λ and the description of the label space L, and outputs a secret key sk and a public key pk. The public key pk implicitly defines a message space M and a set F of admissible functions.

Auth(sk, L, m) :   The authentication algorithm takes a secret key sk, a multi-label $L = (\Delta, \tau)$, and a message $m \in M$ as input and outputs an authenticator σ.

Ver(vk, P_Δ, m, σ) :   The verification algorithm takes as input a verification key vk, a message $m \in M$, an authenticator σ, and a multi-labeled program $P_\Delta = ((f, \tau_1, \ldots, \tau_n), \Delta)$. For verifiable computing schemes based on MACs vk is the secret key sk, for signature based verifiable computing schemes it is the public key pk. The description of the multi-labeled program contains function $f \in F$ taking n inputs, a label $\tau_i \in \{0, 1\}^*$ for each input i for $i = 1, \ldots, n$, and the data set identifier $\Delta \in \{0, 1\}^*$. It outputs '1' if σ is a valid authenticator for m under P_Δ and '0' otherwise.

Eval(pk, $P_\Delta, (\sigma_1, \ldots, \sigma_n)$) :   The evaluation algorithm takes as input the public key pk, a multi-labeled program $P_\Delta = ((f, \tau_1, \ldots, \tau_n), \Delta)$, and a vector of authenticators $(\sigma_1, \ldots, \sigma_n)$ of length n (assuming f takes n inputs). It outputs a new authenticator σ.

The main difference between *verifiable computing schemes based on MACs* and *signature based verifiable computing schemes* is that for the former schemes secret key sk is needed for verification while in the latter verification is performed with the public key pk. For the properties this means that schemes using MACs only offer private verifiability while signature based schemes offer public verifiability.

Definition 5.2 (Authentication Correctness [1]) Homomorphic authenticators satisfy authentication correctness if for any message $m \in M$, all keys (sk, pk) $\leftarrow$ KeyGen($1^\lambda, L$), any multi-label $L = (\Delta, \tau) \in \{0, 1\}^* \times \{0, 1\}^*$, and any authenticator $\sigma \leftarrow$ Auth(sk, L, m), we have that

$$\Pr[\text{Ver}(\text{sk}, Id_L, m, \sigma) = 1] = 1,$$

where Id_L is the identity program with respect to L.

Definition 5.3 (Evaluation Correctness) We fix keys $(\mathsf{sk}, \mathsf{pk}) \leftarrow \mathsf{KeyGen}(1^\lambda, L)$, a function $f : M^n \to M$, and any set of (message, program, tag) triples $\{m_i, P_{\Delta,i}, \sigma_i\}_{i=1}^n$ such that all multi-labeled programs $P_{\Delta,i} = (P_i, \Delta)$ share the same data set identifier Δ and $\mathsf{Ver}(\mathsf{vk}, P_{\Delta,i}, m_i, \sigma_i) = 1$, where $\mathsf{vk} = \mathsf{sk}$ or $\mathsf{vk} = \mathsf{pk}$. If $m = f(m_1, \ldots, m_n), P = f(P_1, \ldots, P_n)$, and $\sigma = \mathsf{Eval}(\mathsf{pk}, P_\Delta, (\sigma_1, \ldots, \sigma_n))$, then

$$\Pr[\mathsf{Ver}(\mathsf{vk}, P_\Delta, m, \sigma) = 1] = 1.$$

To formally define security we look at the following security experiment (due to [1]).

Setup. The challenger generates $(\mathsf{sk}, \mathsf{pk}) \leftarrow KeyGen(1^\lambda, L)$ and gives pk to the adversary A.

Authentication Queries. The adversary can adaptively ask for tags on multi-labels and messages of its choice. Given a query (L, m) where $L = (\Delta, \tau)$, the challenger proceeds as follows: If (L, m) is the first query with data set identifier Δ, then the challenger initializes an empty list $T_\Delta = \emptyset$ for the data set identifier Δ. If T_Δ does not contain a tuple $(\tau, \cdot)$ (i.e. the multi-label (Δ, τ) was never queried), the challenger computes $\sigma \leftarrow \mathsf{Auth}(\mathsf{sk}, L, m)$, returns σ to A and updates the list $T_\Delta \leftarrow T_\Delta \cup (\tau, m)$. If $(\tau, m) \in T_\Delta$ (i.e. the query was previously made), then the challenger replies with the same tag generated before. If T_Δ already contains a tuple for label τ, i.e. (τ, m'), for some $m \neq m'$ then the challenger ignores the query.

Verification Queries. The adversary has access to a verification oracle as follows: Given a query (P_Δ, m, σ) from A, the challenger replies with the output of $\mathsf{Ver}(\mathsf{vk}, P_\Delta, m, \sigma)$, where $\mathsf{vk} = \mathsf{sk}$ or $\mathsf{vk} = \mathsf{pk}$.

Forgery. The adversary terminates the experiment by sending $(P_{\Delta^*}^*, m^*, \sigma^*)$ for some $P_{\Delta^*}^* = (P^*, \Delta^*)$ and $P^* = (f^*, \tau_1^*, \ldots, \tau_n^*)$ to the challenger. Notice that also during the verification query A sends such tuples to the challenger and asks for verification. Thus, if this query is accepted this allows terminating the experiment successfully.
We say a labeled program $P^* = (f^*, \tau_1^*, \ldots, \tau_n^*)$ is well-defined with regards to T_Δ if one of the following conditions holds.

- There exist messages $m_1, \ldots, m_n$ such that the list T_{Δ^*} contains all tuples $(\tau_1^*, m_1), \ldots, (\tau_n^*, m_n)$. Intuitively, this means that the entire input space of f for the data set Δ^* has been authenticated.
- There exist indices $i \in \{1, \ldots, n\}$ such that $(\tau_i^*, \cdot) \notin T_{\Delta^*}$. This happens when A never asks authentication queries with multi-label (Δ^*, τ_i^*) and the function $f(\cdot)$ outputs the same value for all possible unauthenticated inputs.

The experiment outputs '1' if and only if $\mathsf{Ver}(\mathsf{vk}, P_{\Delta^*}^*, m^*, \sigma^*) = 1$ and one of the following conditions holds:

- *Type 1 Forgery:* no list T_{Δ^*} was created during the game, i.e. no message m has been authenticated with respect to data set identifier Δ^* during the experiment.

- *Type 2 Forgery:* Labeled program P^* is well-defined with regards to T_{Δ^*} and $m^* \neq f^*(\{m_j\}_{(\tau_j,m_j)\in T_{\Delta^*}})$, i.e. m^* is not the correct output of the labeled program P^* when executed on previously authenticated messages $(m_1, \ldots, m_n)$.
- *Type 3 Forgery:* P^* is not well-defined with regards to T_{Δ^*}.

Definition 5.4 (Security) A scheme based on homomorphic authenticators is adaptively secure if any PPT adversary A wins the experiment with probability at most $\mathsf{negl}(\lambda)$. It is (weakly) secure if A wins with probability at most $\mathsf{negl}(\lambda)$ without asking verification queries.

5.2 Verifiable Computing Schemes Based on MACs

Based on these homomorphic MACs several verifiable computing schemes have been proposed. All of them are based on bilinear or multilinear maps and for a definition of the respective assumptions, we refer to Appendix A.

The schemes described below are all privately verifiable and only the work in [7] addresses privacy. Note that there are further homomorphic MACS, which are not mentioned here, e.g. [9], that are not known to allow verification faster than computing the function.

5.2.1 Verifiable Delegation of Computation on Outsourced Data

In [1] Backes et al. construct a homomorphic MAC for arithmetic circuits f of degree 2. It is based on bilinear maps (pairings) and pseudo-random functions with so called *closed-form efficiency*. After a preprocessing stage of complexity $O(|f|)$ the client can verify the correctness in constant time. This paper presents a generic way to turn homomorphic MACS with efficient verification into verifiable computing schemes and thus achieves amortized efficiency. Furthermore, it is secure against adaptive adversaries under the Decision Linear assumption (see Assumption A.1).

5.2.2 Generalized Homomorphic MACs with Efficient Verification

In [11] Zhang and Safavi-Naini generalized the verifiable computing scheme presented in [1]. Using ℓ-linear maps, their homomorphic MAC supports arithmetic circuits of depth ℓ. Using the generic transformation of [1], one can thus obtain a verifiable computing scheme for depth ℓ circuits. It also achieves amortized efficiency, while offering security against adaptive adversaries.

5.2.3 Efficiently Verifiable Computation on Encrypted Data

In [7] Fiore et al. show how to combine the homomorphic MACs of [1] with a FHE scheme to construct a verifiable computing scheme for multivariate polynomials of degree 2 that offers input privacy. They furthermore improve on the efficiency by using a homomorphic hash function. Likewise this scheme achieves amortized efficiency and remains secure against adaptive adversaries.

5.3 Signature Based Verifiable Computing on Linear Functions

For certain classes of functions, it is possible to straightforwardly build a verifiable computing scheme from homomorphic signatures. For linear functions

$$f : \mathbb{F}^{k^N} \to \mathbb{F}^n$$

$$v_1, \ldots v_N \mapsto \sum_{i=1}^{N} c_i v_i$$

linearly homomorphic signatures have been proposed, originally in the context of *network coding* (see [2]). There are linear signature schemes that are not known to allow verification faster than computing the function, e.g. [8], and are therefore not considered in this chapter. Note that without combining this with a suitable homomorphic encryption scheme to encrypt the input data, this construction does not provide input-output privacy.

Setup: C generates the keys $(\mathsf{sk}, \mathsf{pk}) \leftarrow \mathsf{KeyGen}(1^\lambda)$ for a linearly homomorphic signature scheme.

Data Outsourcing: To outsource vectors $v_1, \ldots, v_N$, C first signs $w_i = (e_i, v_i)^T$, where e_i is the ith canonical basis vector of $\mathbb{F}^k$ with regards to some label τ, i.e. $\sigma_i \leftarrow \mathsf{Sign}(\mathsf{sk}, w_i, \tau)$ for $i = 1, \ldots, N$ and sends all (w_i, σ_i) to S.

Delegation: C sends $f = (c_1, \ldots, c_N)$ to S.

Computation: S computes $y = \sum_{i=1}^{N} c_i w_i$ and $\sigma \leftarrow \mathsf{Eval}(f, \tau, v_1, \ldots, v_N, \sigma_1, \ldots, \sigma_N)$ and sends (y, σ) to C.

Verification: C checks whether σ is a valid signature for y and whether the result is of the form $y = (c_1, \ldots c_N, \tilde{y})^T$. If both are true it accepts $\tilde{y}$ as the result.

5.3.1 Programmable Hash Functions Go Private: Constructions and Applications to (Homomorphic) Signatures with Shorter Public Keys

In 2015 Catalano et al. [6] presented the first construction that is signature based and allows to verify faster than computing target function f. This solution is based on bilinear maps and uses asymmetric programmable hash functions (see the paper for a formal definition). These signatures are also secure against adaptive adversaries. It should be noted that the preprocessing stage in this paper is divided into the two algorithms KeyGen and EffVerPrep. However, these two algorithms are still only dependent on f and therefore this construction fits our criteria of amortized efficiency. Their scheme offers input privacy towards the verifier, but not towards the server performing the operations. This scheme is based on the hardness of the 2-DHI assumption (see Assumption A.2), the XDDH assumption (see Assumption A.3), and the FDHI assumption (see Assumption A.4).

5.4 Signature Based Verifiable Computing for Polynomial Functions

A broader class of admissible functions are multivariate polynomials of fixed degree. The following works provide signature schemes for polynomial functions. Furthermore, they sketch how these schemes can be used to support verifiable computing. Note that these schemes do not address input-output privacy.

5.4.1 Homomorphic Signatures with Efficient Verification for Polynomial Functions

Catalano et al. constructed homomorphic signatures in [4] based on multilinear maps (specifically those that satisfy Assumption A.10). Their signatures support arithmetic circuits of fixed depth. By using the techniques of [1], one is able to realize a verifiable computing scheme secure against adaptive adversaries that is efficient in an amortized sense.

5.4.2 Algebraic (Trapdoor) One-Way Functions and Their Applications

Catalano et al. show in [3] how to use OWFs to outsource (multivariate) polynomial evaluations of fixed degree. In particular they construct a OWF based on the RSA assumption (see Assumption A.11) and use this to build a verifiable computing scheme. It is the first one that achieves public verifiability without bilinear maps. The scheme is efficient in an amortized sense and offers adaptive security.

5.5   Signature Based Verifiable Computing Using Homomorphic Encryption

Verifiable computing schemes based on homomorphic signatures or homomorphic MACs do not provide data confidentiality w.r.t. the server. Therefore, Lai et al. show in [10] how to generically construct a verifiable homomorphic encryption (VHE) scheme which allows for verifiable computation on outsourced encrypted data.

The constructed VHE combines a homomorphic encryption (HE) scheme, as defined in Chap. 4, and a so called homomorphic encrypted authenticator (HEA). The latter is basically a homomorphic authenticator, as defined in Sect. 5.1, providing *semantic security*.

To formally define semantic security we look at the following security experiment between a challenger and an adversary A (due to [10]).

Setup The challenger runs $(\mathsf{sk}, \mathsf{pk}) \leftarrow \mathsf{KeyGen}(1^\lambda, L)$, gives pk to A, and initializes a list $T_\Delta = \emptyset$.

Authentication Queries A can adaptively ask the challenger for authenticators of its choice. Given a query (L, m) by A, where $L = (\Delta, \tau)$, the challenger proceeds as follows: If $(L, m) \in T_\Delta$, the challenger computes $\sigma \leftarrow \mathsf{Sign}(\mathsf{sk}, L, m)$. If T_Δ does not contain a tuple (L, m) (i.e., the multi-label (Δ, τ) was never queried), the challenger chooses a fresh multi-label $L = (\Delta, \tau) \in \{0, 1\}^* \times \{0, 1\}^*$, computes $\sigma \leftarrow \mathsf{Sign}(\mathsf{sk}, L, m)$, returns σ to A, and updates the list $T_\Delta \leftarrow T_\Delta \cup (L, m)$.

Challenge The adversary submits a multi-label $L = (\Delta, \tau) \in \{0, 1\}^* \times \{0, 1\}^*$ and two messages $m_0, m_1 \in M$. The challenger selects a random bit $\beta \in \{0, 1\}$, computes $\sigma^* \leftarrow \mathsf{Sign}(\mathsf{sk}, L, m_\beta)$, and sends σ^* to the adversary.

Guess The adversary A outputs its guess $\beta^* \in \{0, 1\}$ for β and wins the game if $\beta = \beta^*$.

The advantage of the adversary in this game is defined as $|\Pr[\beta = \beta^*] - \frac{1}{2}|$ where the probability is taken over the random bits used by the challenger and the adversary.

Definition 5.5 (Homomorphic Encrypted Authenticator (HEA)) A HEA is a scheme using homomorphic authenticators as defined in Sect. 5.1 that is *semantically secure*, i.e. where all probabilistic polynomial time adversaries have at most a negligible advantage in the security game described above.

Definition 5.6 (Verifiable Homomorphic Encryption (VHE)) Let $\mathsf{HE} = (\mathsf{HE.KeyGen}, \mathsf{HE.Encrypt}, \mathsf{HE.Decrypt}, \mathsf{HE.Eval})$ be a homomorphic encryption scheme and let $\mathsf{HEA} = (\mathsf{HEA.KeyGen}, \mathsf{HEA.Auth}, \mathsf{HEA.Ver}, \mathsf{HEA.Eval})$ be a HEA. Then, a *VHE* scheme is a tuple of the following PPT algorithms:

$\mathsf{KeyGen}(1^\lambda, L)$:   This algorithm takes a security parameter λ and the description of the label space L as input and outputs a public key $\mathsf{pk} = (\mathsf{pk}_{\mathsf{HE}}, \mathsf{pk}_{\mathsf{HEA}})$ and a secret key $\mathsf{sk} = (\mathsf{sk}_{\mathsf{HE}}, \mathsf{sk}_{\mathsf{HEA}})$, where $(\mathsf{pk}_{\mathsf{HE}}, sk_{\mathsf{HE}}) \leftarrow \mathsf{HE.KeyGen}(1^\lambda)$ and

$(pk_{HEA}, sk_{HEA}) \leftarrow \mathsf{HEA.KeyGen}(1^\lambda, L)$. The public key pk implicitly defines a message space M and a set F of admissible functions.

$\mathsf{EncSign}(sk_{HEA}, pk_{HE}, L, m)$:   This algorithm takes a secret key sk_{HEA}, a public key pk_{HE}, a multi-label $L = (\Delta, \tau)$, and a message $m \in M$ as input. It runs $c_{HE} \leftarrow \mathsf{HE.Encrypt}(pk_{HE}, m)$ and $c_{HEA} \leftarrow \mathsf{HEA.Auth}(sk_{HEA}, L, m)$ and returns $c = (c_{HE}, c_{HEA})$.

$\mathsf{VerDec}(sk_{HE}, pk_{HEA}, P_\Delta, m, c)$:   This algorithm takes a secret key sk_{HE}, a public key pk_{HEA}, a message $m \in M$, a multi-labeled program $P_\Delta = ((f, \tau_1, \ldots, \tau_n), \Delta)$ with $f \in F$, and a ciphertext c as input. If $\mathsf{HEA.Ver}(pk_{HEA}, P_\Delta, c_{HEA}) = 1$ it runs $m \leftarrow \mathsf{HE.Decrypt}(sk_{HE}, c_{HE})$ and outputs m. It outputs '0' otherwise.

$\mathsf{Eval}(pk, P_\Delta, (c_1, \ldots, c_n))$:   This algorithm takes a public key pk, a program $P_\Delta = ((f, \tau_1, \ldots, \tau_n), \Delta)$ with $f \in F$, and a vector of ciphertexts $(c_1, \ldots, c_n)$ of length n (assuming f takes n inputs). It runs $c_{HE} \leftarrow \mathsf{HE.Eval}(pk_{HE}, f, (c_1, \ldots, c_n))$ and $c_{HEA} \leftarrow \mathsf{HEA.Eval}(pk_{HEA}, P_\Delta, (c_1, \ldots, c_n))$ and outputs the new ciphertext $c = (c_{HE}, c_{HEA})$.

The authors used in their work standard homomorphic signature schemes to build the homomorphic encrypted authenticator. This instantiation has the shortcoming that it does not provide an efficient verification process. However, the construction indicates that a verifiable computing scheme that provides not only privacy, but also amortized efficiency can be built, e.g., using the signature scheme proposed by Catalano et al. [5]. Another important requirement for a successful instantiation, which has not been explicitly mentioned by the authors, is that the homomorphic encryption scheme and the homomorphic encrypted authenticator must be homomorphic over the same message space M. Thus, it should be analysed for which pairs of encryption and signature schemes this is provided.

References

1. M. Backes, D. Fiore, R.M. Reischuk, Verifiable delegation of computation on outsourced data, in *2013 ACM SIGSAC Conference on Computer and Communications Security, CCS'13*, Berlin, 4–8 November 2013, pp. 863–874
2. D. Boneh, D.M. Freeman, J. Katz, B. Waters, Signing a linear subspace: signature schemes for network coding, in *Public Key Cryptography - PKC 2009, 12th International Conference on Practice and Theory in Public Key Cryptography, Proceedings*, Irvine, CA, 18–20 March 2009, pp. 68–87
3. D. Catalano, D. Fiore, R. Gennaro, K. Vamvourellis, Algebraic (trapdoor) one-way functions and their applications, in *TCC* (2013), pp. 680–699
4. D. Catalano, D. Fiore, B. Warinschi, Homomorphic signatures with efficient verification for polynomial functions, in *Advances in Cryptology - CRYPTO 2014 - 34th Annual Cryptology Conference, Proceedings, Part I*, Santa Barbara, CA, 17–21 August 2014, pp. 371–389
5. D. Catalano, A. Marcedone, O. Puglisi, Authenticating computation on groups: new homomorphic primitives and applications, in *Advances in Cryptology - ASIACRYPT 2014 - 20th International Conference on the Theory and Application of Cryptology and Information Security, Proceedings, Part II*, Kaoshiung, 7–11 December 2014, pp. 193–212

6. D. Catalano, D. Fiore, L. Nizzardo, Programmable hash functions go private: constructions and applications to (homomorphic) signatures with shorter public keys, in *Advances in Cryptology - CRYPTO 2015 - 35th Annual Cryptology Conference, Proceedings, Part II*, Santa Barbara, CA, 16–20 August 2015, pp. 254–274

7. D. Fiore, R. Gennaro, V. Pastro, Efficiently verifiable computation on encrypted data, in *Proceedings of the 2014 ACM SIGSAC Conference on Computer and Communications Security*, Scottsdale, AZ, 3–7 November 2014, pp. 844–855

8. D.M. Freeman, Improved security for linearly homomorphic signatures: a generic framework, in *Public Key Cryptography - PKC 2012 - 15th International Conference on Practice and Theory in Public Key Cryptography, Proceedings*, Darmstadt, 21–23 May 2012, pp. 697–714

9. R. Gennaro, D. Wichs, Fully homomorphic message authenticators, in *Advances in Cryptology - ASIACRYPT 2013 - 19th International Conference on the Theory and Application of Cryptology and Information Security, Proceedings, Part II*, Bengaluru, 1–5 December 2013, pp. 301–320

10. J. Lai, R.H. Deng, H. Pang, J. Weng, Verifiable computation on outsourced encrypted data, in *Computer Security - ESORICS 2014 - 19th European Symposium on Research in Computer Security, Proceedings, Part I*, Wroclaw, 7–11 September 2014, pp. 273–291

11. L.F. Zhang, R. Safavi-Naini, Generalized homomorphic MACs with efficient verification, in *ASIAPKC'14, Proceedings of the 2nd ACM Workshop on ASIA Public-Key Cryptography*, Kyoto, 3 June 2014, pp. 3–12

Chapter 6
Verifiable Computing Frameworks from Functional Encryption and Functional Signatures

Abstract In addition to proof or argument based verifiable computing schemes and constructions that rely on homomorphic encryption or homomorphic authenticators, verifiable computing schemes can also be constructed using functional encryption or functional signatures. Thus, in this chapter we present the verifiable computing schemes using one of these primitives. Functional encryption refers to encryption schemes where ciphertexts can be decrypted only if they fulfill certain requirements. There are basically two approaches that use functional encryption to build a verifiable computing scheme. "Verifiable Computation from Attribute Based Encryption" by Parno et al. uses (key-policy) attribute-based encryption, a specific instantiation of functional encryption, while the approach presented in "Delegatable Homomorphic Encryption with Applications to Secure Outsourcing of Computation" by Barbosa and Farshim is constructed directly from functional encryption schemes. Functional signatures come with a secondary parameterized signing key, in addition to the master signing key, that allows to sign messages, but restricts the signing capabilities to messages in a certain range. This property allows to build verifiable computing schemes as shown by Boyle et al. in "Functional Signatures and Pseudorandom Functions".

6.1 Verifiable Computation from Functional Encryption

In this section we present the verifiable computing schemes that use (key-policy) attribute-based encryption (ABE) or are directly constructed from functional encryption (FE) schemes. Key-policy ABE (KP-ABE) [8, 10] is a rather recent public key encryption paradigm, where a public key is associated to a universe of attributes A and secret keys are associated to boolean functions f. A holder of a secret key corresponding to f can only decrypt a message encrypted with respect to a subset A' of attributes iff $f(A') = 1$. FE [3] is a very generic definition of various types of public key encryption concepts, such as IBE, ABE, and many other classes. Basically, in such schemes secret keys are associated to a function f and given a ciphertext of a message m under the corresponding public key, the holder of a secret key corresponding to f will only learn $f(m)$ during decryption, instead of learning the full plaintext m. Assuming that the plaintext space has an additional structure and in particular plaintexts are pairs of some (public) index and message

© The Author(s) 2017   37
D. Demirel et al., *Privately and Publicly Verifiable Computing Techniques*,
SpringerBriefs in Computer Science, DOI 10.1007/978-3-319-53798-6_6

space, then one can define FE on predicates over the index space and the key space. In doing so, one obtains KP-ABE as a so called predicate encryption (PE) scheme with a public index.

6.1.1 Verifiable Computation from Attribute Based Encryption

In [9] Parno et al. show how to build a publicly verifiable computation scheme from any key-policy ABE scheme for function family F (that is closed under complement). Their construction verifies the correct output of a function $f : \{0, 1\}^n \rightarrow \{0, 1\}$ that can be computed by a polynomial sized boolean formula. They use the fact that a message encrypted under an attribute x can only be decrypted if $f(x) = 1$ holds. One can extend this to functions f with outputs of arbitrary bit-length by decomposing f into boolean subfunctions $f_1, \ldots, f_n$. The client's computation is independent of f. This approach does not provide input-output privacy and the security has not been analysed yet. ABE schemes fine tailored for this application have been proposed, for instance in [5] and [11].

6.1.2 Delegatable Homomorphic Encryption with Applications to Secure Outsourcing of Computation

Barbosa and Farshim showed in [2] how to create a verifiable computing scheme from a FE scheme, a FHE scheme, and a special type of MACs denoted as *MACs with chameleon keys*. For the relevant definitions and properties of their construction we refer to the original paper. By combining these primitives, this scheme achieves amortized efficiency, while offering public verifiability, and security against adaptive adversaries. It should however be noted that one of the necessary building blocks for this construction, a so called *predicate encryption (PE) scheme for general predicates*, does not exist to the authors knowledge. Today we have functional encryption for any circuit (e.g. see [7]). We, however, note that no efficient instantiations are known. Moreover, they assume that the auxiliary information is transferred not only authentically but also confidentially from the client to the verifier. Their proposed scheme can however handle functions of arity one that can be expressed as k CNF/DNF (conjunctive/disjunctive normal forms) formulas for fixed k. Note that for $k \geq 3$ constructing such formulas is NP hard (see [6]).

6.2 Verifiable Computation from Functional Signatures

In [4] Boyle et al. introduced the concept of functional signature (FS) schemes. In such a scheme two types of signing keys are used. First, a master signing key msk, which allows to compute signatures on arbitrary messages. Second, there is a set of signing keys sk_f which are parameterized by a particular function f. Each key sk_f restricts the signing capabilities to messages in the range of f, i.e. given any m the key sk_f only allows to produce signatures for $f(m)$. Before discussing the application of FS to verifiable computing, we briefly introduce the concept of FS.

Definition 6.1 (Functional Signature (FS) Scheme [4]) A functional signature (FS) scheme for a message space M and function family $F = \{f : D_f \to M\}$ consists of the following polynomial time algorithms:

Setup(1^λ) : The setup algorithm takes as input the security parameter λ and outputs the master signing key msk and master verification key mvk.

KeyGen(msk, f) : The key generation algorithm takes as input the master signing key msk and a function $f \in F$ (represented as a circuit) and outputs a signing key sk_f for f.

Sign(sk_f, f, m) : The signing algorithm takes as input the signing key sk_f, a function $f \in F$, and a message $m \in D_f$ and outputs $f(m)$ and a signature σ for $f(m)$.

Verify(mvk, m, σ) : The verification algorithm takes as input a master verification key mvk, a message m and a signature σ and outputs '1' if the signature is valid and '0' otherwise.

An FS scheme needs to provide the usual correctness property as well as unforgeability. Unforgeability is defined with respect to adaptively chosen signing keys for functions and adaptive signature queries. It requires that under such queries it is infeasible to produce a valid signature for a message that is outside the range of the queried functions and is not the image of any function and message queried to the signing oracle (cf. [4] for formal definitions). Additionally FS schemes may provide the properties of function privacy and succinctness. Informally, the former means that the distributions of signatures on a message generated via different signing keys are computationally indistinguishable and the latter means that the signature size is independent of the size of the message m as well as the size of the description of the function f. In addition to FS as described in [4], in [1] Backes et al. introduced a variation of functional signatures called *delegatable functional signatures*.

6.2.1 Functional Signatures and Pseudorandom Functions

In [4] the authors propose three generic constructions. The first is a naive construction and just requires an adaptively secure (EUF-CMA secure) signature scheme. However, it does neither achieve function privacy nor succinctness.

The second construction uses the first one, but additionally requires a zero-knowledge succinct non-interactive argument of knowledge (SNARK) system in order to achieve function privacy and succinctness. Finally, the third construction drops the succinctness requirement but still preserves function privacy. This is achieved using a non-interactive zero-knowledge arguments of knowledge (NIZKAoK) system instead of SNARKs. Unfortunately, neither of these three construction can be considered practically efficient.

We now sketch the application of FS to verifiable computing, where we align our description with the general definition of a verifiable computing scheme (cf. Definition 2.1). Therefore, let

$$(\mathsf{FS.Setup}, \mathsf{FS.KeyGen}, \mathsf{FS.Sign}, \mathsf{FS.Verify})$$

be a secure, i.e., correct and unforgeable, FS scheme.

$\mathsf{KeyGen}(1^\lambda, f)$:   Based on security parameter λ, run $(\mathsf{msk}, \mathsf{mvk}) \leftarrow \mathsf{FS.Setup}(1^\lambda)$. Set the evaluation key $\mathsf{ek} := \mathsf{sk}_{f'}$ with $\mathsf{sk}_{f'} \leftarrow \mathsf{FS.KeyGen}(\mathsf{msk}, f')$ where $f'(x) := f(x) || x$. The algorithm sets the verification key $\mathsf{vk} := \mathsf{mvk}$ and $\mathsf{sk} := \perp$ and returns $(\mathsf{sk}, \mathsf{vk}, \mathsf{ek})$.

$\mathsf{ProbGen}(\mathsf{sk}, x)$:   The problem generation algorithm does not need to do any preprocessing. It sets $\sigma_x := x$ and $\rho_x := x$. The value σ_x is given to server S while the decoding value ρ_x is kept by client C (but could be made public).

$\mathsf{Compute}(\mathsf{ek}, \sigma_x)$:   Using the evaluation key $\mathsf{ek} := \mathsf{sk}_{f'}$ and the (encoded) input $\sigma_x := x$, S computes and returns an encoded version $\sigma_y := (y, \sigma)$ with $(\cdot, \sigma) \leftarrow \mathsf{FS.Sign}(\mathsf{sk}_{f'}, f', x)$ and $y := f(x)$.

$\mathsf{Verify}(\mathsf{vk}, \rho_x, \sigma_y)$:   Using the verification key $\mathsf{vk} := \mathsf{mvk}$, the decoding value $\rho_x := x$ and the encoded result $\sigma_y := (y, \sigma)$ the verification algorithm computes $b \leftarrow \mathsf{FS.Verify}(\mathsf{mvk}, y||x, \sigma)$ and if $b = 1$ it outputs y and $\perp$ otherwise.

The correctness of this construction follows from the correctness of the FS scheme. Moreover, it is obvious that the verifiable computing scheme obtained is a publicly verifiable computing scheme according to Definition 2.4.

The above construction provides security in the non-adaptive model (weakly secure). Moreover, it is clear that it does trivially not provide input privacy. The efficiency of the above construction is directly related to the underlying FS scheme. In particular, the runtime of the verification equals the runtime of *FS.Verify* and the proof size is equal to the size of the signature of the FS scheme. As the definition of FS does not put any restriction on the time it requires to verify a signature (apart from being polynomial in the security parameter) it depends on the concrete FS scheme used in the construction if the efficiency definitions for verifiable computing are satisfied.

References

1. M. Backes, S. Meiser, D. Schröder, Delegatable functional signatures, in *Public-Key Cryptography - PKC 2016 - 19th IACR International Conference on Practice and Theory in Public-Key Cryptography, Proceedings, Part I*, Taipei, 6–9 March 2016, pp. 357–386
2. M. Barbosa, P. Farshim, Delegatable homomorphic encryption with applications to secure outsourcing of computation, in *Topics in Cryptology - CT-RSA 2012 - The Cryptographers' Track at the RSA Conference 2012, Proceedings*, San Francisco, CA, 27 February–2 March 2012, pp. 296–312
3. D. Boneh, A. Sahai, B. Waters, Functional encryption: definitions and challenges, in *Theory of Cryptography - 8th Theory of Cryptography Conference, TCC 2011* (2011), pp. 253–273
4. E. Boyle, S. Goldwasser, I. Ivan, Functional signatures and pseudorandom functions, in *Public-Key Cryptography - PKC 2014 - 17th International Conference on Practice and Theory in Public-Key Cryptography, Proceedings*, Buenos Aires, 26–28 March 2014, pp. 501–519
5. J. Chen, H. Wee, Semi-adaptive attribute-based encryption and improved delegation for Boolean formula, in *Security and Cryptography for Networks - 9th International Conference, SCN 2014, Proceedings*, Amalfi, 3–5 September 2014, pp. 277–297
6. S.A. Cook, The complexity of theorem-proving procedures, in *Proceedings of the 3rd Annual ACM Symposium on Theory of Computing*, Shaker Heights, OH, 3–5 May 1971, pp. 151–158
7. S. Garg, C. Gentry, S. Halevi, M. Raykova, A. Sahai, B. Waters, Candidate indistinguishability obfuscation and functional encryption for all circuits, in *54th Annual IEEE Symposium on Foundations of Computer Science, FOCS 2013*, Berkeley, CA, 26–29 October 2013, pp. 40–49
8. V. Goyal, O. Pandey, A. Sahai, B. Waters, Attribute-based encryption for fine-grained access control of encrypted data, in *Proceedings of the 13th ACM Conference on Computer and Communications Security, CCS 2006* (2006), pp. 89–98
9. B. Parno, M. Raykova, V. Vaikuntanathan, How to delegate and verify in public: verifiable computation from attribute-based encryption, in *Theory of Cryptography - 9th Theory of Cryptography Conference, TCC 2012, Proceedings*, Taormina, 19–21 March 2012, pp. 422–439
10. A. Sahai, B. Waters, Fuzzy identity-based encryption, in *Advances in Cryptology - EUROCRYPT 2005* (2005), pp. 457–473
11. K. Zhang, J. Gong, S. Tang, J. Chen, X. Li, H. Qian, Z. Cao, Practical and efficient attribute-based encryption with constant-size ciphertexts in outsourced verifiable computation, in *Proceedings of the 11th ACM on Asia Conference on Computer and Communications Security, AsiaCCS 2016*, Xi'an, 30 May–3 June, 2016, pp. 269–279

Chapter 7
Verifiable Computing for Specific Applications

Abstract Beyond the families of schemes we have seen so far, there exist verifiable computing schemes for specific functions, which we present here. More precisely, "From Secrecy to Soundness: Efficient Verification via Secure Computation" by Applebaum et al. allows the computation of arithmetic branching programs, "Signatures of Correct Computation" by Papamanthou et al. allows to compute multivariate polynomials of fixed degree and derivations of multivariate polynomials, "Efficient Techniques for Publicly Verifiable Delegation of Computation" by Elkhiyaoui et al. allows the verification of matrix vector multiplications and univariate polynomials, "Efficient Computation Outsourcing for Inverting a Class of Homomorphic Functions" by Zhang et al. provides verification for the inversion of a class of functions, "Secure Delegation of Elliptic-Curve Pairing" by Chevallier-Mames et al. allows to verifiably compute cryptographic bilinear maps, "Efficiently Verifiable Computation on Encrypted Data" by Fiore et al. presents a way to verify univariate polynomial evaluations over encrypted data, "TrueSet: Nearly Practical Verifiable Set Computations" by Kosba et al. allows to verify *set operations*, "Verifiable Delegation of Computation over Large Datasets" by Benabbas et al. addresses verifiable computing schemes for multivariate polynomials of fixed degree, and "Batch Verifiable Computation with Public Verifiability for Outsourcing Polynomials and Matrix Computations" by Sun et al. provides a batch verifiable computation scheme for multiple functions evaluated on a fixed input.

7.1 From Secrecy to Soundness: Efficient Verification via Secure Computation

In [2] Applebaum et al. show how to use *randomized encodings* together with cryptographic MACs to obtain a verifiable computing scheme. A randomized encoding (RE) for a function F is a non-interactive protocol in which the client uses its randomness r and its input x to compute a message $\hat{y} = \hat{F}(x, r)$ and sends it to the server, who responds by applying a decoder algorithm to $\hat{y}$ to recover $F(x)$. For a detailed description see for example [1].

This yields a privately verifiable computing scheme for *arithmetic branching programs* (see [3]) which only offers security against a weak adversary. It offers input privacy with regard to the server.

© The Author(s) 2017

D. Demirel et al., *Privately and Publicly Verifiable Computing Techniques*,
SpringerBriefs in Computer Science, DOI 10.1007/978-3-319-53798-6_7

7.2   Signatures of Correct Computation

Papamanthou et al. present in [12] the first, and to our knowledge only, framework
for signatures of correct computation (SCC), which implies publicly verifiable
computing. In particular they constructed two SCC schemes, one for multivariate
polynomials of fixed degree d and one for computing the derivations of multivariate
polynomials.

For multivariate polynomials f they used the fact that one can always write

$$f(x_1, \ldots, x_n) - f(a_1, \ldots, a_n) = \sum_{i=1}^{n} (x_i - a_i) q_i(x_1, \ldots, x_n),$$

where $f, q_i \in \mathbb{F}[x_1, \ldots, x_n]$ and the a_i are fixed inputs for $i = 1, \ldots, n$. Working
over a symmetric bilinear group generated by g with bilinear map (or pairing) e
one can compute $\mathsf{FK}(f) = g^{f(t_1, \ldots, t_n)}$ for some random t_i and $i = 1, \ldots, n$. The
server evaluates the function for the given input $a_1, \ldots, a_n$ and writes $f(x_1, \ldots, x_n) -$
$f(a_1, \ldots, a_n)$ like above. It computes $w_i = g^{q_i(t_1, \ldots, t_n)}$ for all $i = 1, \ldots, n$ and gives
the w_i as well as the claimed result v to the client. The client can then check, whether

$$e(\mathsf{FK}(f) \cdot g^{-v}, g) = \prod_{i=1}^{n} e(g^{t_i - a_i}, w_i)$$

holds. If it does, it accepts the result.

Using similar techniques they also construct a scheme to verify the computation
of derivations. Both schemes are set in the standard model, offer adaptive security,
but do not address privacy.

7.3   Efficient Techniques for Publicly Verifiable Delegation of Computation

In [8] two different verifiable computing schemes are presented, one for univariate
polynomials as well as a scheme for matrix-vector multiplication. The former
scheme is unforgeable under the q-strong Diffie-Hellman assumption while the
latter is unforgeable under the co-computational Diffie-Hellman assumption.

Both schemes offer public verifiability and adaptive security but do not address
input-output privacy.

7.4   Efficient Computation Outsourcing for Inverting a Class of Homomorphic Functions

In [14] Zhang et al. present a scheme for verifying the inversion of a class of functions, namely group homomorphisms ϕ where computing a preimage under ϕ is computationally much more expensive than evaluating ϕ. In this case the server's evaluation of ϕ^{-1} can efficiently be verified by computing ϕ.

This scheme offers only private verifiability, but is secure against adaptive adversaries. It also does not depend on any computational assumption and thus provides security in an information-theoretic sense. However, it does not provide privacy.

7.5   Secure Delegation of Elliptic-Curve Pairing

A further scheme for outsourcing a concrete function, in this case a cryptographic bilinear map e, was introduced by Chevallier-Mames et al. in [7]. To compute $e(A, B)$ the client asks the server to compute

$$a_1 = e(A + g_1 G_1, G_2)$$

$$a_2 = e(G_1, B + g_2 G_2)$$

$$a_3 = e(A + g_1 G_1, B + g_2 G_2)$$

$$a_4 = e(s_1 A + r_1 G_1, s_2 B + r_2 G_2)$$

with random points G_1, G_2 and random integers $g_1, g_2, r_1, r_2, s_1, s_2$. The client computes

$$e_{AB} = a_1^{-g_2} \cdot a_2^{-g_1} \cdot a_3 \cdot e(G_1, G_2)^{g_1 g_2}$$

and accepts the result as correct if

$$a_4 = (e_{AB})^{s_1 s_2} \cdot a_1^{r_2 s_1} \cdot a_2^{r_1 s_2} \cdot e(G_1, G_2)^{r_1 r_2 - g_1 r_2 s_1 - g_2 r_1 s_2}$$

holds.

This scheme provides private verifiability, is efficient, and offers input-output privacy. In addition, it is unconditionally secure, so in particular secure against an adaptive adversary.

7.6   Efficiently Verifiable Computation on Encrypted Data

In [9] Fiore et al. present a way to verify univariate polynomial evaluations over encrypted data. The resulting scheme offers private verifiability, input privacy, and adaptive security while providing amortized efficiency.

7.7   Verifiable Delegation of Computation over Large Datasets

In [4] Benabbas et al. presented a verifiable computing scheme for multivariate polynomials of fixed degree d. They propose several instantiations of their scheme of varying efficiency based on different assumptions. Their respective security is based on the hardness of generalized Diffie Hellman assumptions over composite order bilinear maps (see Assumption A.8). Their scheme allows amortized efficiency while offering security against adaptive adversaries. It does not consider privacy.

7.8   Batch Verifiable Computation with Public Verifiability for Outsourcing Polynomials and Matrix Computations

In [13] a *batch verifiable computation scheme* is proposed by Sun et al. This deals with the scenario of multiple functions evaluated on a fixed input. The authors propose schemes for three different function classes: multivariate polynomials of fixed degree, multivariate polynomials of fixed degree in each variable, and matrix vector multiplications.

Their solutions offer security against adaptive adversaries under the Decision Linear assumption (see Assumption A.1) and the co-CDH assumption (see Assumption A.5). They offer public verifiability. Privacy is not considered.

7.9   TrueSet: Nearly Practical Verifiable Set Computations

In [11] Kosba et al. presented a system named TrueSet that allows to verify *set operations*. This scheme supports *set circuits* built on union, intersection, and set difference gates. They presented a variant of Gennaro et al.'s [10] QAPs called *quadratic polynomial programs* (QPP).

It achieves amortized efficiency and decreases the server's overhead by a factor of more than 150 compared to [10]. This scheme does not provide privacy.

Note that works like [5], [6], or [15] which consider multiple clients or servers are beyond the scope of this work.

References

1. B. Applebaum, Randomly encoding functions: a new cryptographic paradigm - (invited talk), in *Information Theoretic Security - 5th International Conference, ICITS 2011, Proceedings*, Amsterdam, 21–24 May 2011, pp. 25–31

2. B. Applebaum, Y. Ishai, E. Kushilevitz, From secrecy to soundness: efficient verification via secure computation, in *Automata, Languages and Programming, 37th International Colloquium, ICALP 2010, Proceedings, Part I*, Bordeaux, 6–10 July 2010, pp. 152–163

3. A. Beimel, A. Gál, On arithmetic branching programs. J. Comput. Syst. Sci. **59**, 195–220 (1999)

4. Z. Brakerski, C. Gentry, V. Vaikuntanathan, Fully homomorphic encryption without bootstrapping. Electron. Colloq. Comput. Complex. **18**, 111 (2011)

5. R. Canetti, B. Riva, G.N. Rothblum, Two protocols for delegation of computation, in *Information Theoretic Security - 6th International Conference, ICITS 2012, Proceedings*, Montreal, QC, 15–17 August 2012, pp. 37–61

6. X. Chen, W. Susilo, J. Li, D.S. Wong, J. Ma, S. Tang, Q. Tang, Efficient algorithms for secure outsourcing of bilinear pairings. Theor. Comput. Sci. **562**, 112–121 (2015)

7. B. Chevallier-Mames, J. Coron, N. McCullagh, D. Naccache, M. Scott, Secure delegation of elliptic-curve pairing, in *Smart Card Research and Advanced Application, 9th IFIP WG 8.8/11.2 International Conference, CARDIS 2010, Proceedings*, Passau, 14–16 April 2010, pp. 24–35

8. K. Elkhiyaoui, M. Önen, M. Azraoui, R. Molva, Efficient techniques for publicly verifiable delegation of computation, in *Proceedings of the 11th ACM on Asia Conference on Computer and Communications Security, AsiaCCS 2016*, Xi'an, 30 May–3 June 2016, pp. 119–128

9. D. Fiore, R. Gennaro, V. Pastro, Efficiently verifiable computation on encrypted data, in *Proceedings of the 2014 ACM SIGSAC Conference on Computer and Communications Security*, Scottsdale, AZ, 3–7 November 2014, pp. 844–855

10. R. Gennaro, C. Gentry, B. Parno, M. Raykova, Quadratic span programs and succinct NIZKs without PCPs, in *Advances in Cryptology - EUROCRYPT 2013, 32nd Annual International Conference on the Theory and Applications of Cryptographic Techniques, Proceedings*, Athens, 26–30 May 2013, pp. 626–645

11. A.E. Kosba, D. Papadopoulos, C. Papamanthou, M.F. Sayed, E. Shi, N. Triandopoulos, TRUESET: faster verifiable set computations, in *Proceedings of the 23rd USENIX Security Symposium*, San Diego, CA, 20–22 August 2014, pp. 765–780

12. C. Papamanthou, E. Shi, R. Tamassia, Signatures of correct computation, in *TCC* (2013), pp. 222–242

13. Y. Sun, Y. Yu, X. Li, K. Zhang, H. Qian, Y. Zhou, Batch verifiable computation with public verifiability for outsourcing polynomials and matrix computations, in *Information Security and Privacy - 21st Australasian Conference, ACISP 2016, Proceedings, Part I*, Melbourne, VIC, 4–6 July 2016, pp. 293–309

14. F. Zhang, X. Ma, S. Liu, Efficient computation outsourcing for inverting a class of homomorphic functions. Inf. Sci. **286**, 19–28 (2014)

15. L.F. Zhang, R. Safavi-Naini, X.W. Liu, Verifiable local computation on distributed data, in *Proceedings of the Second International Workshop on Security in Cloud Computing, SCC@ASIACCS '14*, Kyoto, 3 June 2014, pp. 3–10

Chapter 8
Analysis of the State of the Art

Abstract In this chapter, all verifiable computing schemes discussed in this survey are summarized and their properties are highlighted. We first summarize for each type of verifiable computing scheme presented in the survey, i.e. proof and argument based verifiable computing, verifiable computing from fully homomorphic encryption, homomorphic authenticators, verifiable computing frameworks from functional encryption and functional signatures, and verifiable computing for specific applications, which properties they provide. Like in the rest of the survey the properties concerned are the level of security the scheme provides, how efficient the verification process is, whether anyone or only the client can check the correctness of the result, which function class the verifiable computing scheme supports, and whether privacy with respect to the input and/or output data is given. Afterwards, we discuss to what extent the schemes provide long-term privacy, i.e. are secure against attackers with unbounded computation power. Finally, we discuss for which approaches implementations are available.

8.1 Security, Privacy, and Efficiency

The first property examined is which *function class* the scheme supports. Some support (subsets of) arithmetic circuits, while others can also deal with stateful operations or general loops, i.e. without needing to know the length of the loop during the preprocessing stage. Furthermore, we specify which type of *adversary* the solution can cope with. Some schemes are secure against a strong adversary (S), some are only secure against a weak adversary (W), and for some approaches the security level has not been analysed yet (∅). In addition, we show which *primitives* the construction relies on, since most of them come with further assumptions regarding security. Furthermore, in some scenarios it might be preferable that the scheme provides a certain level of *privacy*. Depending on the type of data, a scheme may either ensure input privacy (I), output privacy (O), input-output privacy (I/O), or no privacy at all (×) with regard to the server. For computing schemes that are publicly verifiable we also highlight whether they provide privacy towards the verifier, i.e. the public. Thus, for a publicly verifiable computing scheme we further distinguish whether it provides input privacy (I), output privacy (O), input-output privacy (I/O), or no privacy at all (×) with respect to the verifier. To be successfully

D. Demirel et al., *Privately and Publicly Verifiable Computing Techniques*,
SpringerBriefs in Computer Science, DOI 10.1007/978-3-319-53798-6_8

Table 8.1 Used
abbreviations

Category	Abbreviation	Explanation
Adversary	S	Strong adversary
	W	Weak adversary
Privacy	I	Input privacy
	O	Output privacy
	I/O	Input-output privacy
	×	No privacy
	NA	No information available
Efficiency	E	Efficient
	A	Amortized efficient
General	D	Dependent on primitives

applied in practice, a scheme also needs to provide *efficiency*. We define a verifiable computing scheme as efficient (E), if the time required for preprocessing and verification is $o(T)$, where T is the time required to compute the function. If only the verification can be performed in $o(T)$, then the computing scheme only provides amortized efficiency (A). Note that verifiable computing schemes that do not provide any of these two types of efficiency have not been discussed in this work. Finally, most solutions are tailored to *private verification*, i.e. where the verification can only be performed by the data owner. However, in some scenarios the verification must be performed by a party different from the owner, requiring the scheme to be *publicly verifiable*. Sometimes it is not possible to make a general statement about a scheme's attributes as they are dependent on choices of primitives (D). The abbreviations introduced here are summarized in Table 8.1.

As shown by Table 8.2 the only proof or argument based approach that provides an efficient generation and verification process is the one proposed in [34]. This scheme, however, only supports a very restricted class of circuits. The other PCP or linear PCP based constructions support larger classes of programs, but achieve only amortized efficiency. In addition, all these approaches are interactive, i.e. require multi round interaction between the server and the client. To reduce the server's overhead later solutions are non-interactive. The latest proposal [3] even achieves input privacy towards the verifier and provides public verifiability. However, all non-interactive proof based schemes use QAPs and are therefore based on *non-falsifiable* assumptions of knowledge, i.e. assumptions that cannot be efficiently denied. As shown in [19] it is actually impossible to build a SNARG (e.g. using QAPs) that is based solely on falsifiable assumptions. This raises some questions on the security of these schemes. For some of these schemes security against the strong adversary has not been investigated. Also note that the schemes offering public verifiability often provide input privacy as an optional feature and can also be used in the standard configuration for data that does not need to remain private.

Constructions based on fully homomorphic encryption naturally offer input-output privacy, because the inputs and correspondingly the outputs are encrypted. However, they do not provide public verifiability. Furthermore, as shown in

Table 8.2 Proof and argument based verifiable computation schemes

Scheme	Function class	A	PrS	E	PV	PrV
[34]/ [33]	Circuits of polylog. depth	S	×	E	✓	NA
[35]	Arithm. circuits	S	×	A	✓	NA
[28]	Arithm. circuits	NA	×	A	×	×
[29]	Arithm circuits + more	NA	×	A	×	×
[30]	Arithm. circuits + more	NA	×	A	×	×
[9]	Stateful	NA	×	A	×	×
[37]	Arithm. circuits	NA	×	A	×	×
[27]	Arithm. circuits + more	S	×	A	✓	I
[15]	Arithm. circuits + more	S	×	A	✓	I
[5]	General loops	S	×	A	✓	I
[6]	General loops	S	×	A	✓	I
[36]	General loops	S	×	A	✓	I
[3]	Arithm. circuits	S	×	A	✓	I

Properties: privacy w.r.t. server (**PrS**), efficiency (**E**), public verifiability (**PV**), privacy w.r.t. verifier (**PrV**)

Table 8.3 FHE based verifiable computation schemes

Scheme	Function class	A	P	E	PV
[18]	Arithm. circuits	W	I/O	A	×
[14]	Arithm. circuits	W	I/O	A	×
[32]	Arithm. circuits	W	I/O	A	×

Property: adversary (*A*), privacy (**P**), efficiency (**E**), public verifiability (**PV**)

Table 8.4 Authenticator based verifiable computation schemes

Scheme	Function class	A	PrS	Primitives	E	PV	PrV
[2]	Poly. of degree 2	S	×	Bilinear maps	A	×	×
[38]	Poly. of fixed degree	S	×	Multilinear maps	A	×	×
[17]	Poly of degree 2	S	I	Bilinear maps	A	×	×
[12]	Linear	S	×	Bilinear maps	A	✓	I
[11]	Poly of fixed degree	S	×	Multilinear maps	A	✓	NA
[10]	Poly of fixed degree	S	×	RSA	A	✓	NA
[24]	D	D	I/O	HE/HEA	D	D	D

Properties: adversary (*A*), privacy w.r.t. server (**PrS**), efficiency (**E**), public verifiability (**PV**), privacy w.r.t. verifier (**PrV**)

Table 8.3 all constructions available are proven secure against a weak adversary only and provide amortized efficiency. Note that currently FHE cannot be considered a practical tool. Thus, how to build efficient solutions that are secure against strong adversaries is still an open question.

The schemes using homomorphic authentication, see Table 8.4, are more restrictive with respect to the supported function class. Furthermore, all schemes only provide amortized efficiency and only the solutions using homomorphic signature schemes provide public verifiability. The generic construction proposed by Lai et al. [24] allows to combine authentication based verifiability with encryption gaining a verifiable computing scheme preserving input-output privacy towards the server. Nevertheless, the function class, security, and efficiency depend on the underlying primitives and further research is required for identifying promising instantiations for different applications.

Another line of research are verifiable computing schemes based on functional encryption or functional signatures, see Table 8.5. The authors of [26], for instance, introduced a primitive built on attribute based encryption. However, for this construction the security has not been analysed yet. The other FE based approach introduced by Barbosa and Farshim [4] is generic, but requires FE for general predicates and can thus not be considered practical yet. Another very interesting approach is to build verifiable computing schemes from functional signatures. Also in this direction more research can be done to allow for a scheme that is efficient and secure against strong adversaries.

Besides the main research directions, i.e. proof and argument based, FHE based, authenticator based, and functional encryption/signature based verifiable schemes, there are also several solutions for specific applications, see Table 8.6. If their properties meet the requirements of the application to be implemented these constructions can also be considered.

Table 8.5 FE and FS based verifiable computation schemes

Scheme	Function class	A	PrS	Primitives	E	PV	PrV
[26]	Boolean functions	$\emptyset$	×	ABE	A	×	×
[4]	D	S	I/O	FE,MAC,FHE,PE	A	✓	D
[7]	Arithm. circuits	W	×	FS	D	✓	NA

Properties: adversary (A), privacy w.r.t. server (**PrS**), efficiency (**E**), public verifiability (**PV**), privacy w.r.t. verifier (**PrV**)

Table 8.6 Other verifiable computation schemes

Scheme	Function class	A	PrV	Primitives	E	PV	PrV
[1]	Arithmetic branching programs	W	I	Randomized encodings	A	×	×
[25]	Poly. + derivations	S	×	Bilinear maps	A	✓	NA
[16]	Univariate polynomials	S	×	Bilinear maps	A	✓	NA
[16]	Matrix-vector multiplication	S	×	Bilinear maps	A	✓	NA
[39]	Inversions of homomorphisms	S	×	$\emptyset$	A	×	×
[13]	Bilinear maps	S	×	Bilinear maps	A	×	×
[17]	Univariate poly.	S	I	Bilinear maps	A	×	×
[23]	Set operations	S	×	QPP	A	✓	NA
[8]	Poly of fixed degree	S	×	Bilinear map	A	×	×
[31]	Multivariate polynomials	S	×	Bilinear maps	A	✓	NA
[31]	Matrix-vector multiplication	S	×	Bilinear maps	A	✓	NA

Properties: adversary (A), privacy w.r.t. server (**PrS**), efficiency (**E**), public verifiability (**PV**), privacy w.r.t. verifier (**PrV**)

The summary shows that the only verifiable computing scheme that achieves efficiency over a single instantiation is a proof based solution. In addition, this line of research has produced constructions that support the most general classes of functions, e.g. general loops and stateful operations. On the downside, their security has not been proven yet, some solutions rely on non-falsifiable assumptions, and privacy towards the server performing the computations is not addressed. Thus, if a verifiable computing scheme providing input-output privacy towards the server for a wide class of functions is needed, one currently has to rely on inefficient approaches using fully homomorphic encryption. Their additional shortcomings are that they are only proven secure against the weak adversary and that they come with an expensive preprocessing phase. However, for use cases that allow the termination of a protocol as soon as one single input is rejected and that do not require an efficient preprocessing phase one may still be interested in using solutions from this line of research. However, before these constructions can be considered for practice more efficient FHE schemes must be developed. Verifiable computing schemes that are authenticator based are proven secure against an adaptive adversary and some solutions even provide public verifiability and/or input-output privacy. On the downside they are currently very restricted with respect to the function class provided. Note however that the operations supported include a huge amount of statistical operations and can therefore be of interest for many concrete instantiations. Furthermore, with respect to the approach providing input-output privacy towards the server, more research has to be done regarding an instantiation gaining (amortized) efficiency. Apart from the numerous solutions based on proofs, arguments, homomorphic encryption, and homomorphic authentication, also other promising approaches, e.g. based on functional encryption, based on functional signatures, and tailored to specific applications, have been proposed. The properties of the construction using functional signatures, for instance, depend on the signature scheme used. Thus, with developing efficient functional signatures also the potential of this approach will increase.

8.2 Long-Term Privacy

For a verifiable computing scheme providing privacy it is an interesting question under which assumptions the input and/or output of the function computed remain secure. On the one hand, a verifiable computing scheme can provide *information theoretical (i.e. statistical or perfect) hidingness*. In this case not even a computationally unbounded attacker, i.e. server or verifier, can violate privacy why these schemes provide so called long-term (or everlasting) privacy. On the other hand, there are schemes that offer only computationally hidingness. Here privacy depends on a cryptographic hardness assumption and is lost as soon as this assumption can be broken for the parameters chosen. With respect to privacy towards the verifier currently the only *perfectly hiding* approach is the homomorphic signature based verifiable computing scheme of [12]. Here the verifier learns nothing about the input data processed from the data published for auditing, i.e. the data processed by the server can be any input with the same probability. The SNARK based approaches of [5, 6, 15, 27], and [3] still provide *statistical hidingness* of the input data with respect to the audit data published. Here the distributions for the possible input values are not exactly the same, but their statistical difference is a negligible function. This is a sufficient condition to achieve long-term privacy. Regarding privacy towards the server, no solution that offers information theoretical privacy is available yet.

8.3 Implementations

So far implementations have only been provided for proof and argument based approaches (see Chap. 3). They can be found in [20–22]. This allows to perform efficiency analyses and to compare the individual solutions. However, for the other types of verifiable computing schemes relying on other primitives, e.g. FHE, authenticators, functional encryption/signatures, corresponding implementations are missing. Thus, implementing these approaches and comparing the efficiency also between different types of verifiable computing schemes would be an interesting task for future work. Furthermore, for most solutions the server's overhead has not been analysed yet. Note that for many applications, such as cloud computing, not only the overhead with respect to the verifier and the client, but also with respect to the server is an important factor to select the most suitable solution. Thus, it would also be very valuable to measure the efficiency of all algorithms that are part of the different verifiable computing approaches.

References

1. B. Applebaum, Y. Ishai, E. Kushilevitz, From secrecy to soundness: efficient verification via secure computation, in *Automata, Languages and Programming, 37th International Colloquium, ICALP 2010, Proceedings, Part I*, Bordeaux, 6–10 July 2010, pp. 152–163
2. M. Backes, D. Fiore, R.M. Reischuk, Verifiable delegation of computation on outsourced data, in *2013 ACM SIGSAC Conference on Computer and Communications Security, CCS'13*, Berlin, 4–8 November 2013, pp. 863–874
3. M. Backes, M. Barbosa, D. Fiore, R.M. Reischuk, ADSNARK: nearly practical and privacy-preserving proofs on authenticated data, in *2015 IEEE Symposium on Security and Privacy, SP 2015*, San Jose, CA, 17–21 May 2015, pp. 271–286
4. M. Barbosa, P. Farshim, Delegatable homomorphic encryption with applications to secure outsourcing of computation, in *Topics in Cryptology - CT-RSA 2012 - The Cryptographers' Track at the RSA Conference 2012, Proceedings*, San Francisco, CA, 27 February–2 March 2012, pp. 296–312
5. E. Ben-Sasson, A. Chiesa, D. Genkin, E. Tromer, M. Virza, SNARKs for C: verifying program executions succinctly and in zero knowledge, in *Advances in Cryptology - CRYPTO 2013 - 33rd Annual Cryptology Conference, Proceedings, Part II*, Santa Barbara, CA, 18–22 August 2013, pp. 90–108

6. E. Ben-Sasson, A. Chiesa, E. Tromer, M. Virza, Succinct non-interactive zero knowledge for a von Neumann architecture, in *Proceedings of the 23rd USENIX Security Symposium*, San Diego, CA, 20–22 August 2014, pp. 781–796
7. E. Boyle, S. Goldwasser, I. Ivan, Functional signatures and pseudorandom functions, in *Public-Key Cryptography - PKC 2014 - 17th International Conference on Practice and Theory in Public-Key Cryptography, Proceedings*, Buenos Aires, 26–28 March 2014, pp. 501–519
8. Z. Brakerski, C. Gentry, V. Vaikuntanathan, Fully homomorphic encryption without bootstrapping. Electron. Colloq. Comput. Complex. **18**, 111 (2011)
9. B. Braun, A.J. Feldman, Z. Ren, S.T.V. Setty, A.J. Blumberg, M. Walfish, Verifying computations with state, in *ACM SIGOPS 24th Symposium on Operating Systems Principles, SOSP '13*, Farmington, PA, 3–6 November 2013, pp. 341–357
10. D. Catalano, D. Fiore, R. Gennaro, K. Vamvourellis, Algebraic (trapdoor) one-way functions and their applications, in *TCC* (2013), pp. 680–699
11. D. Catalano, D. Fiore, B. Warinschi, Homomorphic signatures with efficient verification for polynomial functions, in *Advances in Cryptology - CRYPTO 2014 - 34th Annual Cryptology Conference, Proceedings, Part I*, Santa Barbara, CA, 17–21 August 2014, pp. 371–389
12. D. Catalano, D. Fiore, L. Nizzardo, Programmable hash functions go private: constructions and applications to (homomorphic) signatures with shorter public keys, in *Advances in Cryptology - CRYPTO 2015 - 35th Annual Cryptology Conference, Proceedings, Part II*, Santa Barbara, CA, 16–20 August 2015, pp. 254–274
13. B. Chevallier-Mames, J. Coron, N. McCullagh, D. Naccache, M. Scott, Secure delegation of elliptic-curve pairing, in *Smart Card Research and Advanced Application, 9th IFIP WG 8.8/11.2 International Conference, CARDIS 2010, Proceedings*, Passau, 14–16 April 2010, pp. 24–35
14. K. Chung, Y.T. Kalai, S.P. Vadhan, Improved delegation of computation using fully homomorphic encryption, in *Advances in Cryptology - CRYPTO 2010, 30th Annual Cryptology Conference, Proceedings*, Santa Barbara, CA, 15–19 August 2010, pp. 483–501
15. C. Costello, C. Fournet, J. Howell, M. Kohlweiss, B. Kreuter, M. Naehrig, B. Parno, S. Zahur, Geppetto: versatile verifiable computation, in *2015 IEEE Symposium on Security and Privacy, SP 2015*, San Jose, CA, 17–21 May 2015, pp. 253–270
16. K. Elkhiyaoui, M. Önen, M. Azraoui, R. Molva, Efficient techniques for publicly verifiable delegation of computation, in *Proceedings of the 11th ACM on Asia Conference on Computer and Communications Security, AsiaCCS 2016*, Xi'an, 30 May–3 June 2016, pp. 119–128
17. D. Fiore, R. Gennaro, V. Pastro, Efficiently Verifiable computation on encrypted data, in *Proceedings of the 2014 ACM SIGSAC Conference on Computer and Communications Security*, Scottsdale, AZ, 3–7 November 2014, pp. 844–855
18. R. Gennaro, C. Gentry, B. Parno, Non-interactive verifiable computing: outsourcing computation to untrusted workers, in *Advances in Cryptology - CRYPTO 2010, 30th Annual Cryptology Conference, Proceedings*, Santa Barbara, CA, 15–19 August 2010, pp. 465–482
19. C. Gentry, D. Wichs, Separating succinct non-interactive arguments from all falsifiable assumptions, in *Proceedings of the 43rd ACM Symposium on Theory of Computing, STOC 2011*, San Jose, CA, 6–8 June 2011, pp. 99–108
20. http://cs.utexas.edu/pepper. Retrieved 18 Apr 2016
21. http://research.microsoft.com/verifcomp/. Retrieved 18 Apr 2016
22. https://github.com/scipr-lab/libsnark. Retrieved 18 Apr 2016
23. A.E. Kosba, D. Papadopoulos, C. Papamanthou, M.F. Sayed, E. Shi, N. Triandopoulos, TRUESET: faster verifiable set computations, in *Proceedings of the 23rd USENIX Security Symposium*, San Diego, CA, 20–22 August 2014, pp. 765–780
24. J. Lai, R.H. Deng, H. Pang, J. Weng, Verifiable computation on outsourced encrypted data, in *Computer Security - ESORICS 2014 - 19th European Symposium on Research in Computer Security, Proceedings, Part I*, Wroclaw, 7–11 September 2014, pp. 273–291
25. C. Papamanthou, E. Shi, R. Tamassia, Signatures of correct computation, in *TCC* (2013), pp. 222–242

26. B. Parno, M. Raykova, V. Vaikuntanathan, How to delegate and verify in public: verifiable computation from attribute-based encryption, in *Theory of Cryptography - 9th Theory of Cryptography Conference, TCC 2012, Proceedings*, Taormina, 19–21 March 2012, pp. 422–439
27. B. Parno, J. Howell, C. Gentry, M. Raykova, Pinocchio: nearly practical verifiable computation, in *2013 IEEE Symposium on Security and Privacy, SP 2013*, Berkeley, CA, 19–22 May 2013, pp. 238–252
28. S.T.V. Setty, R. McPherson, A.J. Blumberg, M. Walfish, Making argument systems for outsourced computation practical (sometimes), in *19th Annual Network and Distributed System Security Symposium, NDSS 2012*, San Diego, CA, 5–8 February 2012
29. S.T.V. Setty, V. Vu, N. Panpalia, B. Braun, A.J. Blumberg, M. Walfish, Taking proof-based verified computation a few steps closer to practicality, in *Proceedings of the 21th USENIX Security Symposium*, Bellevue, WA, 8–10 August 2012, pp. 253–268
30. S.T.V. Setty, B. Braun, V. Vu, A.J. Blumberg, B. Parno, M. Walfish, Resolving the conflict between generality and plausibility in verified computation, in *Eighth Eurosys Conference 2013, EuroSys '13*, Prague, 14–17 April 2013, pp. 71–84
31. Y. Sun, Y. Yu, X. Li, K. Zhang, H. Qian, Y. Zhou, Batch verifiable computation with public verifiability for outsourcing polynomials and matrix computations, in *Information Security and Privacy - 21st Australasian Conference, ACISP 2016, Proceedings, Part I*, Melbourne, VIC, 4–6 July 2016, pp. 293–309
32. C. Tang, Y. Chen, Efficient non-interactive verifiable outsourced computation for arbitrary functions. IACR Cryptology ePrint Archive (2014), p. 439
33. J. Thaler, Time-optimal interactive proofs for circuit evaluation, in *Advances in Cryptology - CRYPTO 2013 - 33rd Annual Cryptology Conference, Proceedings, Part II*, Santa Barbara, CA, 18–22 August 2013, pp. 71–89
34. J. Thaler, M. Roberts, M. Mitzenmacher, H. Pfister, Verifiable computation with massively parallel interactive proofs, in *4th USENIX Workshop on Hot Topics in Cloud Computing, HotCloud'12*, Boston, MA, 12–13 June 2012
35. V. Vu, S.T.V. Setty, A.J. Blumberg, M. Walfish, A hybrid architecture for interactive verifiable computation, in *2013 IEEE Symposium on Security and Privacy, SP 2013*, Berkeley, CA, 19–22 May 2013, pp. 223–237
36. R.S. Wahby, S.T.V. Setty, Z. Ren, A.J. Blumberg, M. Walfish, Efficient RAM and control flow in verifiable outsourced computation, in *22nd Annual Network and Distributed System Security Symposium, NDSS 2015*, San Diego, CA, 8–11 February 2015
37. G. Xu, G.T. Amariucai, Y. Guan, Verifiable computation with reduced informational costs and computational costs, in *Computer Security - ESORICS 2014 - 19th European Symposium on Research in Computer Security, Proceedings, Part I*, Wroclaw, 7–11 September 2014, pp. 292–309
38. L.F. Zhang, R. Safavi-Naini, Generalized homomorphic MACs with efficient verification, in *ASIAPKC'14, Proceedings of the 2nd ACM Wookshop on ASIA Public-Key Cryptography*, Kyoto, 3 June 2014, pp. 3–12
39. F. Zhang, X. Ma, S. Liu, Efficient computation outsourcing for inverting a class of homomorphic functions. Inf. Sci. **286**, 19–28 (2014)

Chapter 9
Conclusion

Abstract This work shows that the field of verifiable computing, although not very old, has made huge improvements over the last years. Various solutions have been found for different function classes.

The concrete practicality of these schemes depends on the server's, client's, and verifier's computational overhead, which in turn often depends on the efficiency of the primitives. So advances in fields like FHE, pairings, multilinear maps, circuit generation, or garbled circuits will each be beneficial for the state of the art in verifiable computing. Note that so far there is only one scheme where both, the time required for generation and verification is $o(T)$, where T is the time required to compute the function.

Another requirement that is very important, but only sparely provided in a strong adversary setting, is privacy. There are several attempts to combine verifiable computing schemes secure against adaptive adversaries with privacy preserving ones. There are two constructions available, [1] and [3], that are secure against a strong adversary, provide public verifiability, and potentially provide input/output privacy against both the server computing the output and the public verifying its correctness. For many applications such a primitive would be very valuable. However, concrete instantiations for both solutions are still missing and identifying them would be an interesting task for future work.

There are several publicly verifiable computing schemes offering information theoretically hidingness for the input data processed with respect to the data published for verification, e.g. [2, 4]. However, there are no solutions that provide long-term privacy also towards the server computing the outcome. For applications performing computations on very sensitive data, such as medical records, long-term privacy towards both the server and the verifier would be very valuable. Finally, efficiency has so far not been rigorously studied, e.g. with respect to the server's overhead, leaving open another important task for future work.

This survey shows that, on the one hand, there are many solutions for the task of verifiable computing that provide a bunch of important properties. However, the analysis, on the other hand, also revealed that there are still open issues, e.g.

© The Author(s) 2017

D. Demirel et al., *Privately and Publicly Verifiable Computing Techniques*,
SpringerBriefs in Computer Science, DOI 10.1007/978-3-319-53798-6_9

(long-term) privacy and performance analyses, that need to be addressed before the
existing schemes can be put into practice.

References

1. M. Barbosa, P. Farshim, Delegatable homomorphic encryption with applications to secure
   outsourcing of computation, in *Topics in Cryptology - CT-RSA 2012 - The Cryptographers'
   Track at the RSA Conference 2012, Proceedings*, San Francisco, CA, 27 February–2 March
   2012, pp. 296–312
2. E. Ben-Sasson, A. Chiesa, D. Genkin, E. Tromer, M. Virza, SNARKs for C: verifying program
   executions succinctly and in zero knowledge, in *Advances in Cryptology - CRYPTO 2013 - 33rd
   Annual Cryptology Conference, Proceedings, Part II*, Santa Barbara, CA, 18–22 August 2013,
   pp. 90–108
3. J. Lai, R.H. Deng, H. Pang, J. Weng, Verifiable computation on outsourced encrypted data, in
   *Computer Security - ESORICS 2014 - 19th European Symposium on Research in Computer
   Security, Proceedings, Part I*, Wroclaw, 7–11 September 2014, pp. 273–291
4. B. Parno, J. Howell, C. Gentry, M. Raykova, Pinocchio: nearly practical verifiable computation,
   in *2013 IEEE Symposium on Security and Privacy, SP 2013*, Berkeley, CA, 19–22 May 2013,
   pp. 238–252

Appendix A
Assumptions

Definition A.1 (Bilinear Map [4]) A *bilinear group* is a tuple $(p, \mathbb{G}_1, \mathbb{G}_2, \mathbb{G}_T, e)$ such that:

- $\mathbb{G}_1, \mathbb{G}_2$, and $\mathbb{G}_T$ are cyclic groups of prime order p.
- $e : \mathbb{G}_1 \times \mathbb{G}_2 \rightarrow \mathbb{G}_T$ is bilinear, i.e. for all $g_1 \in \mathbb{G}_1, g_2 \in \mathbb{G}_2$, and $a, b \in \mathbb{Z}$, $e(g_1{}^a, g_2{}^b) = e(g_1, g_2)^{ab}$.
- e is an *admissible* bilinear map, i.e.

  - e is efficiently computable;
  - if g_1 and g_2 are generators of $\mathbb{G}_1$ and $\mathbb{G}_2$, i.e. $\langle g_1 \rangle = \mathbb{G}_1$ and $\langle g_2 \rangle = \mathbb{G}_2$, then $\mathbb{G}_T$ is generated by $e(g_1, g_2)$

- the *Discrete Logarithm Problem* is hard to be computed in $\mathbb{G}_1, \mathbb{G}_2$, and $\mathbb{G}_T$.

The function e is called *bilinear map*, or *pairing*.

Bilinear maps can be divided into symmetric maps where $\mathbb{G} = \mathbb{G}_1 = \mathbb{G}_2$ and asymmetric maps where an isomorphism between $\mathbb{G}_1$ and $\mathbb{G}_2$ is not efficiently computable. There are different hardness assumptions that can be based on bilinear maps.

Assumption A.1 (Decision Linear [1]) Let $(p, \mathbb{G}, \mathbb{G}_T, e) \leftarrow Gen(1^\lambda)$ output a cyclic bilinear group of order p with bilinear map
$e : \mathbb{G} \times \mathbb{G} \rightarrow \mathbb{G}_T$.
The decision linear assumption holds if for randomly chosen $g_b, g_1, g_2 \in \mathbb{G}$ and $r_0, r_1, r_2 \in \mathbb{Z}_p$

$$| \Pr\left[A((p, \mathbb{G}, \mathbb{G}_T, e) \leftarrow Gen(1^\lambda)), g_0, g_1, g_2, g_1^{r_1}, g_2^{r_2}, g_0^{r_1 + r_2}) \right] -$$

$$\Pr\left[A((p, \mathbb{G}, \mathbb{G}_T, e) \leftarrow Gen(1^\lambda)), g_0, g_1, g_2, g_1^{r_1}, g_2^{r_2}, g_0^{r_0}) \right] | = \mathsf{negl}(\lambda).$$

© The Author(s) 2017

D. Demirel et al., *Privately and Publicly Verifiable Computing Techniques*,
SpringerBriefs in Computer Science, DOI 10.1007/978-3-319-53798-6

Assumption A.2 (q-**Strong Diffie-Hellman Inversion** [2]) Let $(p, \mathbb{G}_1, \mathbb{G}_2, \mathbb{G}_T, e) \leftarrow$ $Gen(1^\lambda)$ output an asymmetric bilinear map with random generators $g_1 \in \mathbb{G}_1$ and $g_2 \in \mathbb{G}_2$ and let $q = poly(\lambda)$. The q-Strong Diffie-Hellman Inversion (q-DHI) holds, if for every PPT adversary A

$$\Pr\left[A(g_1, g_1^z, g_2^z, \ldots, g_1^{z^q}, g_2^{z^q}) = g_1^{\frac{1}{z}} \;\middle|\; z \xleftarrow{\$} \mathbb{Z}_p \right] = \mathsf{negl}(\lambda).$$

Assumption A.3 (External Decisional Diffie-Hellman in $\mathbb{G}_1$ [7]) Let $(p, \mathbb{G}_1, \mathbb{G}_2, \mathbb{G}_T, e) \leftarrow Gen(1^\lambda)$ output an asymmetric bilinear map with random generators $g_1 \in \mathbb{G}_1$ and $g_2 \in \mathbb{G}_2$. The External Decisional Diffie-Hellman Assumption (XDDH) holds in $\mathbb{G}_1$, if for every PPT adversary A

$$\left| \Pr\left[A(g_1, g_2, g_1^a, g_1^b, g_1^{ab}) = 1 \;\middle|\; a, b \xleftarrow{\$} \mathbb{Z}_p \right] \right.$$
$$\left. - \Pr\left[A(g_1, g_2, g_1^a, g_1^b, g_1^c) = 1 \;\middle|\; a, b, c \xleftarrow{\$} \mathbb{Z}_p \right] \right| = \mathsf{negl}(\lambda).$$

Assumption A.4 (Flexible Diffie-Hellman Inversion [7]) Let $(p, \mathbb{G}_1, \mathbb{G}_2, \mathbb{G}_T, e) \leftarrow$ $Gen(1^\lambda)$ output an asymmetric bilinear map with random generators $g_1 \in \mathbb{G}_1$ and $g_2 \in \mathbb{G}_2$. The Flexible Diffie-Hellman Inversion Assumption (FDHI) holds, if for every PPT adversary A

$$\Pr\left[W \in \mathbb{G}_1 \backslash \{1\}, W' = W^{\frac{1}{z}} : (W, W') \leftarrow A(g_1, g_2, g_2^z, g_2^v, g_1^{\frac{z}{v}}, g_1^r, g_1^{\frac{r}{v}}) \;\middle|\; z, v, r \xleftarrow{\$} \mathbb{Z}_p \right]$$
$$= \mathsf{negl}(\lambda).$$

Assumption A.5 (Co-computational Diffie-Hellman Assumption [8]) Let $(p, \mathbb{G}_1, \mathbb{G}_2, \mathbb{G}_T, e) \leftarrow Gen(1^\lambda)$ output an asymmetric bilinear map with random generators $g_1 \in \mathbb{G}_1$ and $g_2 \in \mathbb{G}_2$. The co-computational Diffie-Hellman assumption (co-CDH) holds if for all PPT adversaries A

$$\Pr\left[g_1^{ab} \leftarrow A(g_1, g_1^a g_2, g_2^b) \;\middle|\; a, b \leftarrow \mathbb{Z}_p \right] = \mathsf{negl}(\lambda)$$

Assumption A.6 (q-**PDH** [9]) The q-power Diffie-Hellman (q-PDH) assumption holds if for all non-uniform PPT adversaries A

$$\Pr\left[y = g^{s^{q+1}} \;\middle|\; \begin{array}{c} (p, \mathbb{G}, \mathbb{G}_T, e) \leftarrow Gen(1^\lambda) \\ g \leftarrow \mathbb{G} \backslash \{1\} \\ s \leftarrow \mathbb{Z}_p^* \\ \sigma \leftarrow (p, \mathbb{G}, \mathbb{G}_T, e, g, g^s, \ldots, g^{s^q}, g^{s^{q+2}}, g^{s^{2q}}) \\ y \leftarrow A(\sigma) \end{array} \right] = \mathsf{negl}(\lambda).$$

Assumption A.7 (q-PKE [9])   The q power of knowledge assumptions holds if for all adversaries A there exists a non-uniform PPT extractor χ_A such that

$$
\Pr\left[
\begin{array}{c}
\hat{c} = c^a \\
\wedge\, c \neq \prod_{i=0}^{q} g^{a_i s^i}
\end{array}
\;\middle|\;
\begin{array}{c}
(p, \mathbb{G}, \mathbb{G}_T, e) \leftarrow Gen(1^\lambda) \\
g \leftarrow \mathbb{G}\backslash\{1\} \\
a, s \leftarrow \mathbb{Z}_p^* \\
\sigma \leftarrow (p, \mathbb{G}, \mathbb{G}_T, e, g, g^s, \ldots, g^{s^q}, g^a, g^{a \cdot s}, \ldots, g^{a \cdot s^q}) \\
(c, \hat{c}, a_0, \ldots, a_q) \leftarrow (A || \chi_A)(\sigma, z)
\end{array}
\right]
$$

$$
= \mathsf{negl}(\lambda)
$$

for any auxiliary information $z \in \{0, 1\}^{poly(\lambda)}$ that is generated independently of a.

Note that $(y, z) \leftarrow (A || \chi_A)(x)$ signifies that on output x adversary A outputs y and that χ_A, given the same input x and A's random tape, produces z.

Assumption A.8 (q-SDH [3])   The q strong Diffie-Hellman (q-SDH) holds if for all ppt adversaries A

$$
\Pr\left[
y = e(g,g)^{\frac{1}{s+c}}, c \in \mathbb{Z}_p^*
\;\middle|\;
\begin{array}{c}
(p, \mathbb{G}, \mathbb{G}_T, e) \leftarrow Gen(1^\lambda) \\
g \leftarrow \mathbb{G}\backslash\{1\} \\
s \leftarrow \mathbb{Z}_p^* \\
\sigma \leftarrow (p, \mathbb{G}, \mathbb{G}_T, e, g, g^s, \ldots, g^{s^q}) \\
y \leftarrow A(\sigma)
\end{array}
\right] = \mathsf{negl}(\lambda).
$$

For many purposes just having bilinear maps is not enough. Thus, the following is a useful generalization.

Definition A.2 (Multilinear Map [5])   A k-*linear map* is a tuple $pp = (p, \mathbb{G}_1, \ldots, \mathbb{G}_k, e_{ij} : 1 \leq, i, j \leq k)$ such that:

- $e_{ij} : \mathbb{G}_i \times \mathbb{G}_j \rightarrow \mathbb{G}_{i+j}$
- each $(p, \mathbb{G}_i, \mathbb{G}_j, \mathbb{G}_{i+j}, e_{ij})$ is a bilinear map.

There are also several types of assumptions that can be based on multilinear maps.

Assumption A.9 ((k, l) Multilinear Diffie Hellman Inversion, [5])   Let pp (see Definition A.2) be the description of a set of multilinear groups and $g_1 \in \mathbb{G}_1$ a random generator. Let $w \in \mathbb{Z}_p$ be random. The (k, l) Multilinear Diffie Hellman Inversion holds, if for all adversaries A

$$
\Pr\left[A(g_1, g_1^w, \ldots, g_1^{w_l}) = g_k^{w^{kl+1}}\right] = \mathsf{negl}(\lambda).
$$

Assumption A.10 (k Augmented Power Multilinear Diffie Hellman Assumption [6])   Let pp (see Definition A.2) be the description of a set of multilinear groups, and $g_1 \in \mathbb{G}_1$ a random generator. Let $a, b, x \in \mathbb{Z}_p$ be random. The k Augmented

Power Multilinear Diffie Hellman Assumption holds if for all adversaries A

$$\Pr\left[A(g_1, g_1^a, g_1^b, g_1^{ab}, g_1^x, g_1^{ax}, g_1^{abx}) = g_k^{a^{k-1}(bx)^k}\right] = \mathsf{negl}(\lambda).$$

Other well known hardness assumptions are *factorization based*, i.e. given $N = pq$ the product of two distinct odd primes it should be difficult to find p and q. N is called an RSA modulus.

Assumption A.11 (RSA Assumption) Let (N, e) be a pair of integers, where $N = pq$ for some odd prime numbers $p, q, e \in \mathbb{Z}_N \backslash \{1\}$ and $\gcd(e, \varphi(N)) = 1$. Given an element $z \in \mathbb{Z}_N$, the standard RSA problem is to compute the integer y such that $y^e \equiv z \mod N$.

There is also a stronger variant of this where the attacker can choose e.

Assumption A.12 (Strong RSA Assumption) Let (N, e) be a pair of integers, where $N = pq$ for some odd prime numbers $p, q, e \in \mathbb{Z}_N \backslash \{1\}$ and $\gcd(e, \varphi(N)) = 1$. Given an element $z \in \mathbb{Z}_N$ and the freedom to choose e, the *Strong RSA* problem is to compute the integer y such that $y^e \equiv z \mod N$.

References

1. M. Backes, D. Fiore, R.M. Reischuk, Verifiable delegation of computation on outsourced data, in *2013 ACM SIGSAC Conference on Computer and Communications Security, CCS'13*, Berlin, 4–8 November 2013, pp. 863–874
2. D. Boneh, X. Boyen, Efficient selective-ID secure identity-based encryption without random Oracles, in *Advances in Cryptology - EUROCRYPT 2004, International Conference on the Theory and Applications of Cryptographic Techniques, Proceedings*, Interlaken, 2–6 May 2004, pp. 223–238
3. D. Boneh, X. Boyen, Short signatures without random Oracles, in *Advances in Cryptology - EUROCRYPT 2004, International Conference on the Theory and Applications of Cryptographic Techniques, Proceedings*, Interlaken, 2–6 May 2004, pp. 56–73
4. D. Boneh, M.K. Franklin, Identity-based encryption from the Weil pairing. SIAM J. Comput. **32**, 586–615 (2003)
5. D. Catalano, D. Fiore, R. Gennaro, L. Nizzardo, Generalizing homomorphic MACs for arithmetic circuits, in *Public-Key Cryptography - PKC 2014 - 17th International Conference on Practice and Theory in Public-Key Cryptography, Proceedings*, Buenos Aires, 26–28 March 2014, pp. 538–555
6. D. Catalano, D. Fiore, B. Warinschi, Homomorphic signatures with efficient verification for polynomial functions, in *Advances in Cryptology - CRYPTO 2014 - 34th Annual Cryptology Conference, Proceedings, Part I*, Santa Barbara, CA, 17–21 August 2014, pp. 371–389
7. D. Catalano, D. Fiore, L. Nizzardo, Programmable hash functions go private: constructions and applications to (homomorphic) signatures with shorter public keys, in *Advances in Cryptology - CRYPTO 2015 - 35th Annual Cryptology Conference, Proceedings, Part II*, Santa Barbara, CA, 16–20 August 2015, pp. 254–274
8. K. Elkhiyaoui, M. Önen, M. Azraoui, R. Molva, Efficient techniques for publicly verifiable delegation of computation, in *Proceedings of the 11th ACM on Asia Conference on Computer and Communications Security, AsiaCCS 2016*, Xi'an, 30 May–3 June 2016, pp. 119–128

9. B. Parno, J. Howell, C. Gentry, M. Raykova, Pinocchio: nearly practical verifiable computation, in *2013 IEEE Symposium on Security and Privacy, SP 2013*, Berkeley, CA, 19–22 May 2013, pp. 238–252

Printed in the United States
By Bookmasters